A Practical Guide to Age-Period-Cohort Analysis

The Identification Problem and Beyond

A Practical Guide to Age-Period-Cohort Analysis

The Identification Problem and Beyond

Wenjiang Fu

CRC Press
Taylor & Francis Group
Boca Raton London New York

CRC Press is an imprint of the
Taylor & Francis Group, an **informa** business
A CHAPMAN & HALL BOOK

CRC Press
Taylor & Francis Group
6000 Broken Sound Parkway NW, Suite 300
Boca Raton, FL 33487-2742

Printed on acid-free paper
Version Date: 20180405

International Standard Book Number-13: 978-1-466-59265-0 (Hardback)

Visit the Taylor & Francis Web site at
http://www.taylorandfrancis.com

and the CRC Press Web site at
http://www.crcpress.com

To my family, Qi, Martina, and Beverly,

my parents, and my sister Shufen

for all of their love, encouragement and support.

Contents

Preface

Age-period-cohort analysis has received growing attention in recent years thanks to the advances in statistical modeling using cutting edge methods and computational tools. Its broad applications to demography, epidemiology, economics, finance, marketing, public health, and social sciences have helped research in this area thrive with a dramatic increase in the number of publications.

Ever since the early work by Frost (1939) studying mortality from tuberculosis, and the work by Kuhlen (1940) emphasizing the cohort effect on social changes, birth cohort has become an important factor in investigating diseases, human behavior, and related studies. Recognizing the equivalent importance among age, period (calendar year), and cohort (birth cohort) effects, quantitative scientists started an ambitious task of simultaneously estimating all three effects, hoping to disentangle their complex relationship. Over time, this proved to be a very challenging task, evidenced by a large number of publications and numerous approaches that are based on seemingly reasonable assumptions but eventually fail in estimating secular trend with sensible interpretations. An estimation paradox in the statistical models led to a consensus in the late 1980s that the identification problem in age-period-cohort models is unsolvable.

Statistical models and methods have flourished in studying complex high dimensional data in the twenty-first century. Some potentially shed light on the identification problem, including non-parametric smoothing methods and regularization models. These stimulated novel ideas in addressing the identification problem at the prime time, and resulted in new development in statistical theory and computational methods that either directly address the identification problem or provide alternative approaches.

This monograph aims to introduce statistical models and tools in age-period-cohort analysis, and explain why the difficult identification problem can be resolved and how to solve it. It provides a practical guide to practitioners, health professionals, and social science researchers, for exploring age-period-cohort analysis through basic models and graphics, understanding the identification problem, studying the advanced methods in addressing this difficult problem, and learning how to analyze age-period-cohort data with efficient statistical software using user-friendly R programming. To graduate students in statistics and biostatistics, and statisticians who are interested in thoroughly understanding the identification problem, the methods, and the theory, it provides explanations and justification in full detail using statistical finite sample and large sample theories.

Part I of the book first motivates age-period-cohort analysis using real examples and data sets in Chapter 1, and introduces graphical methods and basic models, including single factor and two factor linear and loglinear models in Chapters 2–3. Second, it introduces full age-period-cohort models, illustrates the identification problem with cancer mortality data, and briefly reviews numerous approaches and the challenges in Chapters 4–5. Third, it provides non-technical approaches to the identification problem, and explains why it can be resolved and how to solve it with the intrinsic estimator method in Chapter 6. Results on the related statistical large sample studies and algorithm are explained with no technical details and easy to understand. Fourth, it demonstrates the intrinsic estimator method in data analysis with both linear models and loglinear models in Chapter 7.

Part II of the book provides full technical details of theoretical justification and related matters. First, in Chapter 8, it proves that the intrinsic estimator yields unbiased estimation and converges to true parameter values as the sample size diverges to infinity, while other estimators yield asymptotic bias. Second, it derives the variance of the period and cohort effect estimates for generalized linear models using the Delta method in Chapter 9. Furthermore, it studies the selection of side conditions by comparing the variance of parameter estimates, and confirms the efficiency of the centralization approach. Finally, in Chapter 10, it studies a special class of age-period-cohort data with unequal spans in age and period groups, and develops the intrinsic estimator method in a different setting to address the identification problem without losing information in the original age and period groups.

Part I can be used for advanced undergraduate and master level graduate teaching in quantitative sciences. A background of basic statistical models and mathematical statistics is sufficient, and knowledge in advanced statistics is not required. Part II requires advanced statistical theory at a PhD level in statistics and biostatistics, including maximum likelihood, profile likelihood, penalty models, consistency and asymptotic normality. The exercises at the end of each chapter can be used for homework practice.

During my statistical career, I have learned from many people on this topic and would like to give my sincere thanks. Shelly Bull introduced this topic to me when I was a PhD student. Stephen Fienberg had a long discussion with me on a snowy winter day in his office, where, for the first time, I identified the estimable function on his blackboard, though he was not convinced without seeing a rigorous proof. Peter Hall hosted my visit at ANU in 2003, where we worked together on the asymptotics. He also encouraged me to study the nonparametric approach. Sir David R. Cox wrote me a very enthusiastic letter with full encouragement, and shared his thoughts when I was puzzled in the early years. Christopher Winship had a number of long discussions with me in Boston. Jun Liu and Jiming Jiang helped improve my work through many critical discussions. David Grubbs and Sherry Thomas of CRC provided very helpful editorial assistance. Martina Fu offered professional editing and helped proofread the manuscript. Finally, I would like to thank the grant funding from the NIH/NCI.

Wenjiang Fu
Houston, Texas
January 2018

List of Figures

List of Tables

Part I

Age-Period-Cohort Models, Challenges, Methods, and Rationale

1

Motivation of Age-Period-Cohort Analysis — Examples and Applications

1.1   What Is Age-Period-Cohort Analysis?

Age-period-cohort analysis, or APC analysis for short, is a popular and efficient tool to analyze data of a certain event, such as death, diagnosis of cancer, crime, etc., in order to understand how the likelihood of the event depends on the age, period (calendar year), and cohort (birth cohort), particularly when individual records are not available due to security or confidentiality, or when the number of individual records becomes intractably huge, such as for the U.S. national mortality data.

Age-period-cohort analysis often concerns the secular trend of disease incidence or mortality (in particular, chronic diseases) in epidemiology, or the trend of social events or deaths in social studies or demography. In recent years, applications have also been observed in economics, finance and marketing research to study investment, family income, sales, etc. The public health studies often investigate various types of cancer, cardiovascular diseases, birth defect, diabetes, obesity, neurological disorders, etc. The demography and social studies monitor birth, death, migration, literacy, marriage, belief in religion, violent crime, juvenile delinquency, suicide, etc. The economic studies focus on changes in employment, family income, investment, etc. The marketing research observes consumption, or sales of a product, such as coffee, soft drinks, life insurance policies, etc.

Scientific evidence has shown that the development of diseases (such as cancer) often varies with age due to the growth or aging of the human body, and that human behavior (such as violent crime) varies with age as people learn and mature. Consequently, age is no doubt an important factor in studying human diseases and behavior. The introduction of disease prevention programs (such as cancer prevention programs) and treatments, or intervention to adverse behaviors, is expected to reduce the likelihood of event occurrences, aiming to cure and ultimately eradicate the diseases, or to minimize the violence in a community. The success of such programs needs to be assessed over time, which may often take a number of years, and the period effect of the event occurrence needs to be monitored through the entire duration.

Perhaps the earliest cohort study was proposed by Frost (1939) to study the mortality from tuberculosis. Likewise, the earliest article emphasizing the cohort effect was perhaps Kuhlen (1940) on social change and its impact on psychological studies of life span. Ever since, the importance of cohort effects has been recognized and age-period-cohort studies have been conducted by many health researchers, sociologists, and demographers to investigate how the likelihood of the event occurrence, often quantified by the event rate (e.g. cancer incidence or mortality rate, or suicide rate), varies across a time span in a population. An age-period-cohort study examines the rate or frequency of an event in a population within a geographic region, estimates the effects of the age, period, and birth cohort on the likelihood of the event occurrence, and identifies secular trends in the age, period, and cohort. Such studies not only help policymakers to identify emerging issues to address public health concerns (e.g. diseases, suicide), but also help scientific investigators to raise questions and form hypotheses that may lead to further investigations for an ultimate resolution.

1.2   Why Age-Period-Cohort Analysis?

Before introducing age-period-cohort analysis, let's look at a few examples.

Example 1: A public health study. A public health office would like to investigate whether lung cancer mortality in the United States has decreased over the last forty years given the educational programs of smoking related diseases and smoking cessation during the last couple of decades. The investigator obtained archived lung cancer mortality data among US males and US females aged 20 to 85+ years old from 1980 to 2009. Since the number of deaths from lung cancer is about 150,000 each year among both males and females and the number of newly diagnosed lung cancer cases is about 224,000 each year, it is more efficient to group individual mortality records into summary counts to study the grouped mortality data summarized in a table of 5 year intervals of age groups in rows and 5 year intervals of periods in columns. Tables 1.1 and 1.2 display grouped lung cancer mortality data among males and females, respectively, obtained from the Surveillance Epidemiology and End Results (SEER) cancer registry data in the U.S. In each cell of the table is a mortality rate together with the total number of deaths from lung cancer. The last age group, 85+, includes all ages of 85 years and older, and thus has a span of more than five years. Since the number of deaths from lung cancer among those of 90 years and older is relatively small, one may regard the last age group as a 5 year interval in the study.

Since lung cancer mortality may vary with many factors, including exposure to certain environmental risk factors, such as asbestos or cigarette smoking, it is thus expected that the mortality rate may grow less aggressively, or even decrease, with calendar year after the implementation of stringent regulations on environmental risk exposure and smoking cessation programs in recent years. Age is another important factor in studying mortality as aging puts seniors at a higher risk. In addition, the generation effect also plays an important role in mortality. Older generations had different lifestyles and lower living standards during their early years of life than younger generations. These in turn affect behaviors such as alcohol and tobacco consumption, which is known to have an impact on lung cancer mortality. In order to gain insight into the lung cancer mortality trend over time, the investigator needs to examine the mortality rate with respect to age, period, and cohort to determine 1) which factor among the three influences lung cancer mortality the most, 2) whether men or women in a specific age or cohort group have a higher risk than others so that further studies may focus on these high risk groups for more efficient intervention, and 3) whether lung cancer mortality rate increases, fluctuates, or decreases during the study period so that the effect of treatment and intervention programs can be assessed.

Example 2: A social study. A sociologist is interested in investigating whether the economic depression in recent years has led to increased suicide or crime rates in a metropolitan area. The sociologist collected the numbers of cases of suicides and homicides in the metropolitan area with the age of each individual and year of each incidence from 1980 to 2010 and obtained the total population of the metropolitan area for the same period of time. Since the number of suicides or homicides in each year is not large and may be unstable, it is recommended that the data be clustered into 3 year groups in age and calendar year so that the grouped data will present a more stable trend. It is also known that on average, maturity and behavior vary with age and education, thus the suicide and crime rates may be influenced by age, social activities, overall economy, etc. Hence, the age, period, and cohort are expected to be important factors that influence the suicide rate and crime rate over time.

Example 3: A finance study. An investment banker is interested in learning whether a pattern can be found in how people's attitude towards investment changes when they grow older, and how it changes over time, with a particular focus on the changes around the time of the financial market crash. It is known that, in general, senior people pay more attention to pension and healthcare benefits, but it is less known how economic development influences people's attitude towards investment, and whether older generations invest more in the stock market than younger generations. The investment banker may collect data of individual investors aged 25–70 years from 1980 to 2010, and further study the investment behaviors or types of funds (e.g. high-risk-high-yield or low-risk-low-yield) that people in a specific age group and generation are interested in, and how these change over time when the financial market experiences a sharp downturn. The banker may want to learn how the age, period, and birth cohort influence investment patterns. Due to the large amount of data, a 5 year or 10 year group in age and period may be formed to study the attitude towards investment.

The above examples illustrate that age-period-cohort analysis can be used in many areas to study human behavior, health status, or other events that

TABLE 1.1
Mortality from Lung Cancer among US Males from 1980 to 2009
Rate (per 10^5 person-year) and Frequency

Age	Year					
	1980–84	1985–89	1990–94	1995–99	2000–04	2005–09
20–24	0.1	0.1	0.1	0.1	0.1	0.1
	68.0	48.0	46.0	48.0	55.0	51.0
25–29	0.3	0.3	0.2	0.2	0.2	0.2
	168.0	173.0	128.0	119.0	102.0	128.0
30–34	1.2	1.2	1.1	1.0	0.6	0.6
	535.0	623.0	636.0	515.0	331.0	280.0
35–39	5.3	4.3	4.1	3.3	2.7	1.9
	2026.0	1986.0	2142.0	1898.0	1467.0	983.0
40–44	17.5	14.8	12.1	10.8	9.9	6.9
	5404.0	5616.0	5619.0	5707.0	5581.0	3755.0
45–49	46.9	40.5	33.8	26.1	24.3	20.8
	12734.0	12337.0	12632.0	12036.0	12707.0	11650.0
50–54	99.2	92.2	78.7	61.8	50.9	45.9
	27158.0	24407.0	23528.0	23107.0	23395.0	23679.0
55–59	173.8	172.6	156.0	126.8	105.1	84.8
	47081.0	44851.0	39779.0	36809.0	38232.0	38015.0
60–64	262.5	274.6	264.6	223.9	189.3	155.2
	63983.0	68790.0	64668.0	54657.0	52723.0	54155.0
65–69	367.9	373.5	381.3	343.8	299.4	255.9
	73804.0	81067.0	86010.0	77200.0	67611.0	66597.0
70–74	463.0	482.8	475.1	462.6	418.1	360.7
	69322.0	78507.0	86287.0	89387.0	81319.0	72033.0
75–79	510.9	545.8	549.7	526.1	510.6	464.5
	50216.0	61205.0	69617.0	75591.0	79177.0	73595.0
80–84	498.9	559.7	593.6	571.0	543.2	529.8
	26845.0	34703.0	43208.0	48780.0	54003.0	58719.0
85+	400.9	463.8	520.9	534.6	517.0	506.1
	14371.0	18460.0	23897.0	29533.0	33638.0	40255.0

may be influenced by age, period, and birth cohort simultaneously. In the
remainder of this chapter, I will demonstrate what types of data can be studied
with age-period-cohort analysis, and describe in later chapters how to use

TABLE 1.2

Mortality from Lung Cancer among US Females from 1980 to 2009
Rate (per 10^5 person-year) and Frequency

Age	1980–84	1985–89	1990–94	1995–99	2000–04	2005–09
20–24	0.1	0.1	0.1	0.0	0.1	0.1
	28	33	32	21	30	35
25–29	0.2	0.2	0.2	0.2	0.2	0.2
	95	122	106	97	78	80
30–34	0.8	0.8	0.9	0.8	0.6	0.5
	374	414	477	431	308	223
35–39	3.4	2.7	2.8	3.0	2.6	1.8
	1312	1294	1514	1709	1405	924
40–44	9.8	8.9	7.5	7.7	8.1	6.6
	3152	3498	3557	4123	4686	3647
45–49	24.0	23.1	20.8	16.9	17.2	17.7
	6848	7318	8086	8030	9265	10178
50–55	44.1	47.3	44.8	38.5	32.3	31.9
	12980	13335	14117	15074	15510	17170
55–59	68.7	78.3	81.5	73.8	65.8	53.3
	20843	22389	22566	23026	25474	25378
60–64	92.8	115.0	126.9	124.1	116.7	100.3
	26136	33345	35159	33675	35710	38039
65–69	114.1	145.3	172.1	179.2	176.4	162.5
	28507	38813	47527	47582	45940	47880
70–74	120.2	168.1	207.3	230.9	235.8	227.1
	24885	37408	49842	57802	57166	54655
75–79	110.7	161.8	218.5	249.0	272.3	277.6
	17341	28575	42048	52079	59399	58590
80–84	102.3	141.7	198.7	245.1	274.4	288.3
	10414	16700	26868	36711	45002	49936
85+	92.6	115.6	154.1	191.7	224.3	246.8
	7950	11686	18468	26918	34650	42468

statistical models to identify patterns of secular trends in the age, period, and

birth cohort for a given data set.

1.3   Four Data Sets in APC Studies

I now provide four real data sets with detailed descriptions. The first two are lung cancer mortality data among US males and US females. The third is the US mortality data in both males and females who were diagnosed with the Human Immunodeficiency Virus (HIV). The fourth is the mean value of retirement accounts of the US families in social surveys from 1989 to 2010. These data sets will be used in this book to demonstrate the statistical models and methods in age-period-cohort analysis using graphical and analytic approaches.

Lung Cancer Mortality Data among US Males and US Females

Tables 1.1 and 1.2 display the lung cancer mortality data among US males and US females, respectively. Both have 14 rows and 6 columns with 5 year intervals for age and period, except for the last age group (85+), which has more than 5 years. The 14 rows represent age groups from 20–24 years old to 80–84 years old and 85+. The 6 columns represent period groups from 1980–84 to 2005–09. Since the deaths in the last age group were often predominantly among 85–89 years old, this age group is often regarded as a 5 year age interval in age-period-cohort analysis. Each cell of the table has two numbers, the age–period specific mortality rate of lung cancer (per 100,000 person-years) and the number of deaths. Notice also that the tables have 19 diagonals (from top-left to bottom-right) with the number of cells on each diagonal increasing from 1 on the two extreme diagonals (in the bottom-left and top-right corners) to a maximum of 6 on the center diagonals. These 19 diagonals represent birth cohorts, where, within each cohort, the men or women who died from lung cancer were born in the same birth years. For example, the oldest cohort (in the bottom-left corner) of Table 1.1 were the men who were born in 1899 or earlier, while the next diagonal were born between 1896 and 1904, and the youngest cohort (in the top-right corner) was born between 1981 and 1989. Also note that there is usually an overlap between consecutive cohorts. For example, the above two oldest cohorts have a four year overlap in birth years (1896–1899), although the consecutive age and period intervals have no overlap at all. While this overlap between consecutive cohort intervals has not been utilized by many studies, it can be beneficial to accommodate the overlap in cohort effect modeling, which I will discuss in later chapters. The study data were downloaded from the SEER database and will be analyzed to identify the secular trend of lung cancer mortality.

Mortality Data from Those Who Were Diagnosed with HIV Table 1.3 displays mortality data from US men and women who were diagnosed with human immunodeficiency virus (HIV) from 1987 to 2013. The table has 14 age groups and 6 periods with 5 year intervals for each, except for the last period, which covers only two years (2012–2013), and the last age group with age 80 years and older. Each cell has the mortality rate (per 100,000 person-years) and the frequency (the number of deaths) in the specific age group during the specific years. The fewer number of years in the last period does not change the period and cohort coding, and thus does not need special attention as long as the frequency in the period is not too small to make the rate unstable. The HIV mortality data were obtained from the SEER mortality database. I will discuss the secular trend estimation of the HIV mortality data, similar to the modeling of the lung cancer mortality data.

Mean Value of Retirement Accounts of the US Families 1989–2010 Table 1.4 displays the mean value (in 2010 US dollars) of the retirement

TABLE 1.3

Mortality from HIV among US Males and Females 1987–2013

Rate (per 10^5 person-year) and Frequency

Age	Year					
	1987–1991	1992–1996	1997–2001	2002–2006	2007–2011	2012–2013
20–24	2.7	2.8	1.0	0.7	0.6	0.5
	2593.0	2578.0	904.0	743.0	636.0	211.0
25–29	12.1	15.9	4.0	2.3	1.5	1.2
	13027.0	15832.0	3934.0	2209.0	1589.0	518.0
30–34	21.9	33.9	9.8	5.2	2.8	1.8
	23765.0	37952.0	10238.0	5186.0	2765.0	765.0
35–39	25.0	40.0	13.4	9.7	4.7	2.6
	24430.0	44139.0	15210.0	10298.0	4756.0	1014.0
40–44	20.7	37.4	14.1	12.3	7.2	3.8
	17635.0	36849.0	15627.0	13948.0	7596.0	1577.0
45–49	15.4	26.5	12.1	11.7	8.0	5.4
	10228.0	22420.0	11864.0	12893.0	9080.0	2334.0
50–54	10.3	17.6	8.2	9.5	7.4	5.8
	5787.0	11716.0	6938.0	9338.0	8098.0	2626.0
55–59	6.6	11.4	5.7	6.2	6.0	5.2
	3521.0	6289.0	3685.0	5210.0	5798.0	2198.0
60–64	4.0	7.1	3.9	4.0	4.1	3.9
	2123.0	3639.0	2064.0	2550.0	3303.0	1415.0
65–69	2.3	3.9	2.4	2.9	2.8	2.7
	1139.0	1944.0	1169.0	1467.0	1673.0	777.0
70–74	1.3	2.0	1.5	1.7	1.9	1.8
	520.0	891.0	647.0	721.0	852.0	370.0
75–79	0.9	1.0	0.8	1.0	1.1	1.3
	280.0	320.0	286.0	369.0	419.0	200.0
80–84	0.5	0.5	0.4	0.5	0.7	0.9
	100.0	111.0	93.0	139.0	195.0	100.0
85+	0.4	0.4	0.2	0.3	0.4	0.4
	58.0	70.0	47.0	68.0	115.0	42.0

TABLE 1.4

Mean Value of Retirement Accounts (in 2010 dollar) of US Households 1989–2010

Age	Year							
	1989	1992	1995	1998	2001	2004	2007	2010
< 35	15.4	20.3	26.7	30.1	23.1	29.2	26.2	26.9
35–44	46.8	41.5	50.4	69.7	79.5	78.4	83.9	84.5
45–54	87.8	106.1	123.0	120.6	154.7	163.3	162.2	172.5
55–64	105.0	107.6	132.8	191.1	243.5	249.1	283.5	291.3
65–74	83.9	82.2	116.5	138.0	214.9	240.3	279.8	309.4
75+	55.6	84.3	83.9	123.9	155.5	136.7	110.6	173.8

accounts of US households in a social survey from 1989 to 2010. The table has six rows of age groups for the head of the household: from < 35, 35–44, to 65–74 and 75 +, and eight columns of survey year or period with three years apart: from 1989 to 2010. Since the number of years in each age group differs from that in the periods, the diagonals of the table do not reflect birth cohorts and the cohort effects need to be estimated with a special coding scheme. The use of unequal spans between age and period groups is not uncommon in social survey data and thus needs special attention when modeling age, period, and cohort effects. I will introduce a special coding scheme and models accordingly to address this unequal span issue.

1.3.1 Special Features of These Data Sets

These data sets represent a variety of APC data and will be used in later chapters to demonstrate statistical models and methods, and how to deal with the unique features of each data set, such as unequal spans in the age groups and periods. The first two tables are typical APC data sets. Both have 5 year intervals in age and period, except for the last age group. Each cell of the table has two numbers: a rate and a frequency count of cases in a given age and period. The diagonals form birth cohorts and represent the people who were born in about the same years, allowing estimation of cohort effects. Often, the consecutive cohorts have overlaps, such as 4 year overlaps as shown in Tables 1.1 and 1.2. Table 1.2 has a rate 0 in one cell, which results in missing data following a log-transformation of the rate for linear models.

Table 1.4 has a special feature — unequal spans in age groups and periods. The age groups have 10-year spans while the periods are 3 years apart, or equivalently, have 3-year spans. Such unequal spans make the diagonals invalid in representing true birth cohorts. Furthermore, if one would like to collapse the periods into 9 year groups, close to the 10 year span in the age groups, it would be difficult to estimate secular trend in period as there would be only 3 or at most 4 periods in the collapsed data, leading to a loss of information. I will discuss this unequal span feature, which requires a special method for modeling and trend estimation. Another interesting issue is that a table of single year periods and multiple year age groups seems to have no overlap between consecutive cohorts, but the coding of cohort effects requires collapsing multiple periods into one, resulting in multi-year spans in the periods, and consequently, overlaps between consecutive cohorts.

1.4   Data Source

There are various data sources for APC studies. Seeing as most practitioners have their own data, I provide a few websites here to help graduate students who are interested in learning the analytic skills and applying them to real world problems.

The data source largely depends on the subject area and topic of interest of each study. Most often, health registry data are available in many countries, such as the SEER data, the Centers for Disease Control and Prevention (CDC) mortality data in the U.S., and the Ontario Cancer Registry data of Canada. Sociological data may be obtained from large scale social surveys on birth, death, marriage, workforce, etc. Population data are often made available through census and population surveys by government agencies. Other types of data can be either obtained from local government agencies or through research surveys, such as crime rate data from the local police authority of a municipality, or religious belief data from survey studies of church goers in a community. For the convenience of readers, here I provide websites of government agencies and non-profit organizations that provide data.

1) Surveillance Epidemiology and End Results (SEER) Cancer Registry Data in the US

The SEER data cover more than ten cancer registries in major metropolitan areas in the US, including cancer incidence and mortality by age, year of diagnosis, race of patients, cancer sites, and other variables starting from the year 1970 to the most recent year 2009. It has a user friendly software package SEER*Stat available free of charge at the website http://seer.cancer.gov/.

2) Berkeley Human Mortality Database

The Berkeley Human Mortality Database (HMD) covers social survey data in 37 countries and regions, including Australia, Canada, Chile, Israel, Japan, New Zealand, Taiwan, the U.S., and many European countries. It covers birth and death data up to 2015 in many countries, including life table data of France as early as 1816. The database website is http://www.mortality.org.

3) General Social Survey by the NORC at the University of Chicago

This site has many links to national survey data, including the American National Election Studies from 1948 to 2008, the entire US census microdata from 1990 to 2008, the California census data in 1990 and 2000, the Racial and Prejudice Survey data from 1986 to 1998, the California workforce survey in 2001–2002, a large US population survey from 1962, and the National Health Interview Survey data from 1969. The website is `http://www3.norc.org/GSS+Website`.

4) CDC National Center for Health Statistics

The US Centers for Disease Control and Prevention (CDC) has national health statistics and mortality data by state, race, age, period, and other variables. Its website (`http://www.cdc.gov/nchs/`) provides information about the data and summary statistics, such as the secular trend of age-adjusted rate, etc.

5) Data source and information in other countries

Other sources include the cancer registry of the Province of Ontario, Canada through Cancer Care Ontario (`https://www.cancercare.on.ca/cms/One.aspx?portalId=1377&pageId=121950`), the European Cancer Organisation (`http://www.ecco-org.eu/`), the International Agency for Research on Cancer (`http://www.iarc.fr/en/research-groups/index.php`), and the World Health Organization (`http://www.who.int/research/en/`).

1.5   R Programming and Video Online Instruction

All **R** functions used in this book are provided online. The four data sets are also included for the reader's convenience.

A video instruction of how to install the **R** package and how to use **R** functions to generate plots and graphs of the APC data is provided online. In addition, a video illustrating the complexity of the APC models with computer dynamic graphics is also provided.

1.6 Suggested Readings

The age-period-cohort analysis has been applied to many studies in demography, economics, epidemiology, marketing research, psychology, public health, and sociology. Methodological studies can be traced back to Greenberg et al (1950) and Mason et al (1976). A large number of studies were published during the 1970s and 1980s, exploring statistical models and effect estimates of the age, period, and birth cohort, including Smith et al (1982), Rodgers (1982,1982a), Janis et al (1981, 1985), Kupper et al (1983a, 1983b, 1985a, 1985b), Mason and Entwisle (1985), Holford (1983, 1985, 1991), Fienberg and Mason (1985), Clayton and Schifflers (1987a, 1987b). The number of publications on methodologies declined in the 1990s with the recognition of the difficulty in resolving the identification problem in the statistical model, Lee and Lin (1996, 1996a, 1996b). The publications focused more on the observations on field work, such as disease trends in different types of cancer, Tarone and Chu (1992). Interestingly, following a few papers published in the late 1990s and early 2000s, however, a surge of the publications on age-period-cohort analysis has been observed, reflecting novel approaches to addressing the identification problem using smoothing techniques (Heuer 1997, Fu 2008), regularization method (Fu 2000), the principal component analysis (Fu 2000, Yang et al 2004, 2008, Fukuda 2006, Smith 2008, Fu 2016). In recent years, not only have there been more publications on health investigation, economic research, socio-behavioral studies, but also calls for the resolution to the identification problem have come from both theoretical and practical aspects (Rosenberg and Anderson 2011).

1.7   Exercises

1. Lay out the reasons why the identification problem cannot be resolved by searching through the publications.

2. Find an interesting data set in the form of rows of age groups, and columns of periods that is of interest to your work, and describe it.

3.* Search through the methodological papers, and present all the reasons, if possible, why the identification problem cannot be resolved. Are you convinced by these reasons?

---

* Difficult exercises with an asterisk are meant for graduate students in biostatistics.

2

Preliminary Analysis — Graphic Methods

In this chapter, I introduce a few basic graphic methods to visualize the secular trend of the APC data in age, in period, and in birth cohort, using a user-friendly **R** function for APC plots. The graphic methods have been studied in previous years (e.g. Kupper et al 1983, 1985), and most recently by Robertson and Boyle (1998b). I demonstrate the 2-dimensional (2D) and 3-dimensional (3D) surface plots with the examples of APC data given in Section 1.3.

2.1 2D Plots in Age, Period, and Cohort

Perhaps the simplest and most commonly used procedure in APC analysis is the visualization of the rate data by plotting the rate against the age, period, or cohort. I illustrate the plot using the lung cancer mortality rate data among US males in Table 1.1. To plot the data, I first organize the rates in an $a \times p$ table with the **R** command `apcheader` to add labels for the age and period in the rate matrix. This command will be also used in data analysis later so that the estimates of age, period, and cohort effects in the analysis output will be properly labeled.

```
> x=apcheader(r=USlungCAmale.r, agestart=20, yearstart=1980,
    agespan=5, yearspan=5, header=F)
```

```
> x[1:4,]
```

	1980	1985	1990	1995	2000	2005
20	0.1	0.1	0.1	0.1	0.1	0.1
25	0.3	0.3	0.2	0.2	0.2	0.2
30	1.2	1.2	1.1	1.0	0.6	0.6
35	5.3	4.3	4.1	3.3	2.7	1.9

Generate 2D plot with R function `apcplot`

```
x=USlungCAmale
apcplot(x, header=T, apc="age", xleg=70, yleg=4, scale="log")
```

The options `xleg` and `yleg` specify the location of the legend. The option `scale="log"` is the default, the alternative `scale="lin"` generates the original scale plot of the rate. Similarly, the plot of the rate against period (Figure 2.2) or cohort (Figure 2.3) can also be made using the same function with specified options.

```
apcplot(x, header=T, apc="per", pylim=c(-5,6.5), xleg=1982,
   yleg=-3, nleg=5)
apcplot(x, header=T, apc="coha", xleg=1900, yleg=4)
apcplot(x, header=T, apc="cohp", xleg=1900, yleg=4)
```

In the above commands, the option `apc="coha"` specifies the cohort effect plot by age groups, and `apc="cohp"` specifies the cohort effect plot by period groups.

Figure 2.1 presents the age trend of the rate by period in the original scale and the logarithmic scale. It can be seen that the mortality rate increases with age till 70 in every period. Comparing between the two panels, the curves are not parallel in the original scale (upper panel) but are more or less parallel in

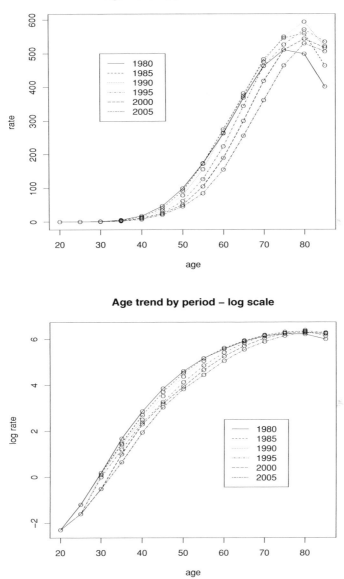

FIGURE 2.1
Plot of lung cancer mortality rate among US males against age by period.
Upper panel: original scale; Lower panel: log scale. The parallel pattern in the
log scale indicates the additive effects.

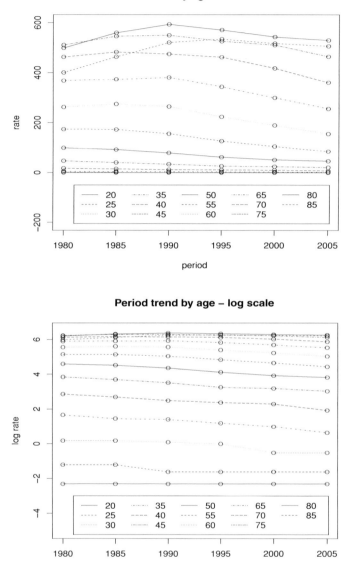

FIGURE 2.2
Plot of lung cancer mortality rate among US males against period by age.
Upper panel: original scale; Lower panel: log scale. The parallel pattern in the
log scale indicates the additive effects.

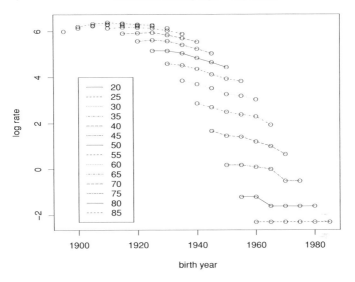

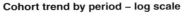

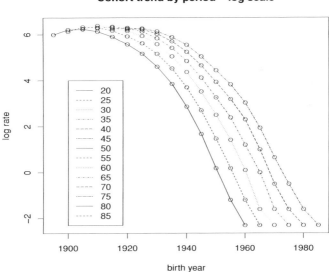

FIGURE 2.3

Plot of lung cancer mortality rate among US males against cohort. Upper panel: cohort trend by age; Lower panel: cohort trend by period.

the logarithmic scale (lower panel). This indicates that after the logarithmic transformation, the mortality rate is additive across different periods (i.e. the age effects and period effects are additive), which translates into multiplicative age and period effects in the original scale. Similarly, parallel patterns are observed in the period trend by age in the logarithmic scale (Figure 2.2), indicating the additive effects of age and period in the logarithmic scale. This helps to explain why taking the logarithmic transformation of the rate has been a common practice in APC analysis.

Figure 2.3 shows plots of the cohort trend by age group and by period group. It is shown that cohort effects are flat within age groups, but decrease sharply within periods. In either way, it can be concluded that the older cohorts had a much higher mortality rate than the younger ones.

It can be inferred from these observations that the disease rate presents interesting and intriguing patterns in the age, period, and cohort. The 2D plot is a simple and important method to explore APC data, and may provide insight to age, period, and cohort effects as well as disease trends. Furthermore, it illustrates the complexity of APC analysis.

2.2   3D Plots in Age, Period, and Cohort

Another graphical method to examine APC data is to plot the 3-dimensional (3D) surface of the rate in the age, period, and cohort. Contrary to the 2D plots, the 3D plot shows how the rate changes with the age, period, and cohort simultaneously. However, it is worthwhile to note that such a plot can only be used for basic description of APC data, as it provides neither rigorous modeling of the data nor a confirmation or justification of any conclusion made out of the data. Thus readers are advised to use caution in interpreting the 3D plot.

Figure 2.4 presents two 3D surface plots of the lung cancer mortality rate among US males: the original scale in the upper plot and the log scale in the lower plot. Different patterns between the two plots are observed, especially in the moderate and old age groups and in the corner of the oldest cohorts. I demonstrate below how to draw 3D plots with the **R** function `apcplot`.

Generate 3D plot with R function `apcplot`

```
x=apcheader(r=USlungCAmale.r, agestart=20, yearstart=1980,
  agespan=5, yearspan=5, header=F)
apcplot(x, header=T, P3D=T, scale="lin", p3dcol=0, theta=-50,
  phi=30)
title (main="3D plot - linear scale")
```

The option `P3D=T` generates 3D plot, while the default `P3D=F` generates 2D plot. The options `theta` and `phi` specify the angles of the plot. The default option `scale="log"` generates log-scale plot, while the alternative `scale="lin"` generates the original scale plot. The option `p3dcol=0` specifies the color of the surface. Notice that in a 3D plot, age and period need to be specified for the x-axis and y-axis, respectively, while the rate is specified for the z-axis.

2.3   Suggested Readings

Early work on plots of APC data can be found in Kupper et al (1983), which provides explanation for fitting loglinear models or taking the logarithmic

3D plot – linear scale

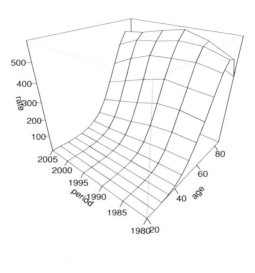

3D plot – log scale

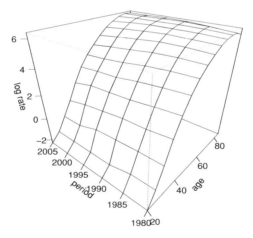

FIGURE 2.4
3D Plots of lung cancer mortality rate among US males against age and period.
Upper panel: original scale; Lower panel: log scale.

transformation of event rates in APC analysis. Robertson and Boyle (1998b) also provides a systematic review of graphical methods, including 2D and 3D plots.

2.4 Exercises

1. Plot age, period, and cohort curves of the lung cancer mortality rate among US females. Compare the trends with the US male mortality rate and explain the differences.

2. Plot the 3D surface of the lung cancer mortality rate among US males with different angles by varying the values of the angles θ (theta) and ϕ (phi). Compare the plots with those in Figure 2.4.

3

Preliminary Analysis of Age-Period-Cohort Data — Basic Models

Having observed the 2D and 3D plots of the APC data, one may wonder how to analyze the data. Specifically, what models and methods are available to estimate the secular trends in the age, period, and cohort for a given data set? Furthermore, what can be concluded based on the secular trend estimation? Is it possible to identify groups of specific age or birth cohort, in which subjects are at higher risk in developing the disease or carrying out the event of interest? To answer these questions, I introduce several statistical models in this chapter. Depending on the type of response variables, two popular models will be discussed: a linear model for continuous Gaussian response of the event rate or measure, and a loglinear model for event frequency count out of population exposure.

Based on the factors included for model covariates, the models can be single factor models, two factor models, or three factor models. Single factor models include age effect (A) model, period effect (P) model, and cohort effect (C) model. Two factor models include age-period (AP) model, age-cohort (AC) model, and period-cohort (PC) model. Three factor models include all age, period, and cohort effects. I introduce the single factor models and two factor models in this chapter, and leave the three factor models to later chapters because of the complexity. I will answer the following questions. Among the single factor and two factor models, which models fit the data well? What criteria can be used to select an appropriate model? These are typical questions in statistical modeling, and the practice is often referred to as model selection. I will introduce model selection and assessment criteria in this chapter.

3.1   Linear Models for Continuous Response

I first consider linear models for APC data. Tables 1.1–1.4 are common in that each cell of the table has a continuous quantity, such as the mortality rate of a disease or a summary value of retirement accounts. In such a case, especially when the response takes values in the continuum of a large range of positive numbers, a linear model of the quantity or its logarithm is of interest. Such a log transformation is often needed and justifiable for the additive effects as discussed in section 2.1 and illustrated in Figures 2.1–2.3. However, exceptions exist, as shown in this book. Furthermore, to make the model flexible with no assumptions on *a priori* patterns of the age, period, or cohort effects, fixed effects of the factors are specified so that the effects will be determined solely by the data rather than by the investigator's subjective assumptions. Popular assumptions, such as a quadratic or cubic age effect, may not be valid for a given data set, and require further testing. Hence, such an assumption approach will not be considered in this book.

Denote by a and p the numbers of age groups and periods, respectively. So the number of diagonals is equal to $a + p - 1$. Notice that the number of diagonals is equal to the number of birth cohorts if the age and period have identical spans as in Tables 1.1 and 1.2. Following the discussion in section 1.3.1, birth cohorts do not follow diagonals of the table when the data have unequal spans in age and period, as shown in Table 1.4. For model simplicity, assume identical spans in the age and period for now. Unequal spans will be considered in later chapters. Let $\alpha_1, \ldots, \alpha_a$ be fixed age effects, $\beta_1, \ldots, \beta_p$ be fixed period effects, and $\gamma_1, \ldots, \gamma_{a+p-1}$ be fixed cohort effects. Furthermore, let μ be the model intercept to represent the overall mean effect. Denote by Y_{ij} the response or its logarithm in cell (i, j), in the i-th age group and j-th period. I study the single-factor models and two-factor models below.

3.1.1 Single Factor Models

Age effect model A single factor age effect model is

$$Y_{ij} = \mu + \alpha_i + \varepsilon_{ij}, \qquad i = 1, \ldots, a, \text{ and } j = 1, \ldots, p, \qquad (3.1)$$

where ε_{ij} are independent random errors and follow a Gaussian distribution with mean 0 and common variance σ^2, i.e. $\varepsilon_{ij} \sim N(0, \sigma^2)$. This one-way analysis of variance (ANOVA) model (3.1) is overparametrized with $a + 1$ parameters $\mu, \alpha_1, \ldots, \alpha_a$. By linear model theory, unique parameter estimates can be achieved by specifying a side condition on the parameters, either the centralization

$$\sum_{i=1}^{a} \alpha_i = 0, \qquad (3.2)$$

or a pre-determined reference level

$$\alpha_I = 0 \text{ for some } 1 \le I \le a. \qquad (3.3)$$

Although a unique set of parameter estimates is achieved by a side condition, the estimates also vary with the side condition, leading to biased estimation, except for a special class of parameters — the contrasts. A linear combination of model parameters $L = c_1 \alpha_1 + \ldots + c_a \alpha_a$ is a contrast if the coefficients $c_1, \ldots, c_a$ sum up to zero. In particular, the difference between any two estimates $\alpha_{i'} - \alpha_{i''}$ is invariant for any $1 \le i' \ne i'' \le a$. It thus implies that the estimates of $(\alpha_1, \ldots, \alpha_a)$ as a collection of parameters remain invariant subject to a translation by a constant, i.e. the relative values of these a parameters remain the same regardless of the coordinates. Consequently, the age trend or the geometric pattern of age effects $\alpha_1, \ldots, \alpha_a$ is uniquely estimated without bias. Since estimating the pattern of age effects rather than the effect values themselves is of primary interest in APC analysis, age effect model (3.1) yields unbiased estimation regardless of side conditions.

Although age trend estimation is invariant to side conditions, the standard errors of parameter estimates vary with side conditions. It has been revealed in a recent work (Fu et al 2017) that among all side conditions, the centralization yields the smallest variance of parameter estimates $\widehat{\alpha}_1, \ldots, \widehat{\alpha}_a$. I will discuss this issue in Chapter 9, and thus refer interested readers to detailed theoretical derivation and simulations there. Hence in this book, I will use parameter centralization for side conditions unless other side conditions are specified.

To estimate the parameters in the age effect model (3.1), I take the least-squares estimates by minimizing the residual sum of squares

$$\min_{(\mu, \alpha_1, \ldots, \alpha_a)} \sum_{i=1}^{a} \sum_{j=1}^{p} (y_{ij} - \mu - \alpha_i)^2,$$

and obtain the estimates $\widehat{\mu} = \bar{Y}_{..}$ and $\widehat{\alpha}_i = \bar{Y}_{i.}$ for $i = 1, \ldots, a$, where the notations follow the same convention as in the ANOVA model, a bar over a quantity (e.g. $\bar{Y}_{i.}$) denotes the average of the quantity across all values of the index replaced with a dot $(\cdot)$. The variance component σ^2 is estimated by $\widehat{\sigma}^2 = \sum_{ij} r_{ij}^2 / (ap - a)$ with model residuals $r_{ij} = Y_{ij} - (\widehat{\mu} + \widehat{\alpha}_i)$. These ANOVA estimates provide unbiased estimation of the overall mean μ, age effects α_i with $i = 1, \ldots, a$ and the variance component σ^2.

Period effect model Similar to age effect model (3.1), one may fit a single factor period effect model

$$Y_{ij} = \mu + \beta_j + \varepsilon_{ij}, \qquad i = 1, \ldots, a, \text{ and } j = 1, \ldots, p, \qquad (3.4)$$

where β_j is the fixed effect of period group j, while the other parameters remain the same as in model (3.1). The model is fitted by the least-squares method, yielding unbiased estimation of the intercept and period effects $\widehat{\mu} = \bar{Y}_{..}$ and $\widehat{\beta}_j = \bar{Y}_{.j}$ for $j = 1, \ldots, p$, similar to the age effect model. Model (3.4) has $ap - p$ degrees of freedom for the variance component in the period effect model, different from age effect model (3.1). The variance component is estimated by $\widehat{\sigma}^2 = \sum_{ij} r_{ij}^2 / (ap - p)$, where the residuals $r_{ij} = y_{ij} - \widehat{\mu} - \widehat{\beta}_j$.

Cohort effect model A cohort effect model, however, differs largely from the above age effect and period effect models. The special form of APC data is balanced for age effect and period effect models, but unbalanced for cohort effect model. It can be seen that the oldest cohort (first diagonal) and the youngest cohort (last diagonal) have only one cell, while the cohorts in the middle have a maximum number of cells on each diagonal, which is equal to $\min(a, p)$. This may potentially lead to a problem in fitting a cohort effect model as discussed below. The cohort effect model takes a form similar to that of the age effect or period effect model.

$$Y_{ij} = \mu + \gamma_k + \varepsilon_{ij}, \qquad i = 1, \ldots, a, \text{ and } j = 1, \ldots, p, \qquad (3.5)$$

where γ_k is the k-th diagonal cohort effect with $k = a - i + j$. To derive the estimates for the mean μ and cohort effect γ_k, I take the least-squares estimates by minimizing the residual sum of squares

$$\min_{(\mu, \gamma_1, \ldots, \gamma_{a+p-1})} \sum_{i=1}^{a} \sum_{j=1}^{p} (y_{ij} - \mu - \gamma_k)^2.$$

For fixed $k = 1, \ldots, a + p - 1$, denote by m_k the number of cells on the k-th diagonal. Setting the partial derivatives to 0 leads to

$$m_k\mu + m_k\gamma_k = \sum_{a-i+j=k} y_{ij} \ .$$

Notice that $\sum_k \gamma_k = 0$ by the parameter centralization. The estimates of the mean and cohort effects are given by

$$\widehat{\mu} = \frac{1}{a+p-1} \sum_{k=1}^{a+p-1} \left(\frac{1}{m_k} \sum_{a-i+j=k} y_{ij} \right)$$

and

$$\widehat{\gamma}_k = \left(\frac{1}{m_k} \sum_{a-i+j=k} y_{ij} \right) - \widehat{\mu} \ ,$$

i.e. the mean μ is estimated with the mean of the cohort effects of all diagonals, and the cohort effect γ_k is estimated with the deviation of the k-th diagonal effect from the overall mean.

Since the degrees of freedom is $ap - (a + p - 1) = (a - 1)(p - 1)$ in the cohort effect model, the variance component is estimated with $\widehat{\sigma}^2 = \frac{1}{(a-1)(p-1)} \sum_{ij} r_{ij}^2$, where $r_{ij} = y_{ij} - \widehat{\mu} - \widehat{\gamma}_k$.

Notice that since APC data is unbalanced for the cohort effect model due to the varying number of observations on the diagonals, the estimate of mean μ in the cohort effect model (3.5) is not the grand mean of all observations as in the age effect and period effect models, but rather the mean of the cohort effects with each cohort effect being estimated by only the observations on the diagonal. The varying number of observations in the estimation of the cohort effect reflects the imbalance of APC data. Ignorance of the imbalance would lead to biased parameter estimation of the cohort effect model.

3.1.2 Two Factor Models

As in ANOVA analysis, single factor models may not fit the data well if the effects of more than one factor are present. So two factor models are also considered, including the age-period (AP) model and age-cohort (AC) model. Although in theory, period-cohort model is also possible and of interest, age effect has been found important in general in APC analysis. Hence, a model with no age effect is not of interest and will be omitted here. In case a period-cohort model is needed for specific data, one may follow the age-cohort model to fit a period-cohort model similarly.

Age-period model An AP model takes the form

$$Y_{ij} = \mu + \alpha_i + \beta_j + \varepsilon_{ij}, \qquad i = 1, \ldots, a, \text{ and } j = 1, \ldots, p, \qquad (3.6)$$

with all parameters described as before. The least-squares method yields unbiased parameter estimation $\widehat{\mu} = \bar{y}_{..}$, $\widehat{\alpha}_i = \bar{y}_{i.} - \bar{y}_{..}$ and $\widehat{\beta}_j = \bar{y}_{.j} - \bar{y}_{..}$, following the conventional notations of ANOVA model. The age and period trend estimates are unbiased regardless of side conditions, following similar discussions to the one in the age effect model. Since the model's number of parameters is $1 + (a-1) + (p-1) = a + p - 1$, the residual degrees of freedom is $(a-1)(p-1)$, which is the same as the cohort effect model. The variance component is estimated with $\widehat{\sigma}^2 = \frac{1}{(a-1)(p-1)} \sum_{ij} r_{ij}^2$, where $r_{ij} = y_{ij} - (\widehat{\mu} + \widehat{\alpha}_i + \widehat{\beta}_j)$.

Age-cohort model An AC model takes the form

$$Y_{ij} = \mu + \alpha_i + \gamma_k + \varepsilon_{ij}, \qquad i = 1, \ldots, a, \text{ and } j = 1, \ldots, p, \qquad (3.7)$$

with all parameters described as before and $k = a - i + j$. Similar to the cohort effect model, APC data is unbalanced for AC model. The parameters can be estimated with the least-squares method

$$\min_{(\mu, \alpha_1, \ldots, \alpha_a, \gamma_1, \ldots, \gamma_{a+p-1})} \sum_{i=1}^{a} \sum_{j=1}^{p} (y_{ij} - \mu - \alpha_i - \gamma_k)^2.$$

Although it is difficult to derive a closed form for the least-squares estimates of the AC model (see exercise at the end of the chapter), it can be estimated with a computer program as shown in section 3.1.3. Notice that the degrees of freedom of the AC model is $2a + p - 2$, larger than the AP model since it requires more parameters. The variance component is estimated with $\widehat{\sigma}^2 = \frac{1}{(a-1)(p-2)} \sum_{ij} r_{ij}^2$, where residuals $r_{ij} = y_{ij} - (\widehat{\mu} + \widehat{\alpha}_i + \widehat{\gamma}_k)$, $\widehat{\mu}$, $\widehat{\alpha}_i$ and $\widehat{\gamma}_k$ are the least-squares estimates, and $k = a - i + j$.

3.1.3   R Programming for Linear Models

In this section, I provide **R** programs for fitting the linear models. First, notice that most statistical analysis software packages use a reference category in estimating fixed effects of categorical variables because the primary task is to make inference on the contrasts. As I discussed briefly in the last section, more accurate estimation can be achieved with the centralization side condition for its smaller variance than the reference level side conditions. Hence, many ANOVA model procedures in various statistical software packages may not fit APC models properly. For example, the SAS procedures use the last category for reference, leading to larger standard errors. I will demonstrate the single factor and two factor models using the centralization side condition. I here provide **R** functions and show how to use these functions to conduct APC analysis. These functions are made available through the book website and can be downloaded for free. Readers can compare the ease and friendliness of these **R** functions with the existing software packages available in other languages, such as SAS and STATA. I use the US lung cancer mortality rate data to demonstrate model fitting. For simplicity, I ignore the unit of the mortality rate (10^{-5}) in fitting the linear model since it only affects the estimate of the model intercept μ but not age, period, and cohort effects, which are of primary interest in APC analysis.

R program to fit a linear model to event rate data

Add header to data set

```
y=apcheader(r=USlungCAmale.r, agestart=20, yearstart=1960,
 agespan=5, yearspan=5, head=F)
```

Fit a single factor model with options to choose model and to plot

```
apclinfit(r=y, apcmodel="A", transform="log", header=T, Plot=T)
apclinfit(r=y, apcmodel="P", transform="log", header=T, Plot=F)
apclinfit(r=y, apcmodel="C", transform="log", header=T, Plot=T)
```

The option `transform="identity"` specifies no transformation on the
response variable, while the log transformation `"log"` is the default.

Fit a two factor model
AP model with plot:

```
apclinfit(r=y, apcmodel="AP", header=T, Plot=T)
```

AC model with plot:

```
apclinfit(r=y, apcmodel="AC", header=T, Plot=T)
```

Tables 3.1, 3.2, 3.4 and 3.5 display the output of model fitting by the **R** function `apclinfit` for the age effect model, period effect model, age-period model, and age-cohort model, respectively. Table 3.3 compares the goodness-of-fit of these single factor and two factor models using R^2 (`R.squared`) and the adjusted R_a^2 (`adjusted.R.squared`). It can be seen that the best model among all those above is the age-cohort model, which has the largest adjusted R-squares. The period effect model has the lowest R-squares (0.5536), indicating a potential lack of fit to the data. The parameter estimates, their standard errors, and the corresponding t-values and p-values are also provided in the tables corresponding to the individual model fit. Figures 3.1 and 3.2 provide plots of effect estimates and the pointwise 95% confidence intervals of the estimates.

TABLE 3.1

Output of Linear Age Effect Model by **R** Function `apclinfit` on Lung Cancer Mortality

```
> apclinfit(r=y, apcmodel="A", header=T, transform="log",
    Plot=TRUE)
$model
lm(formula = log(rr) ~ x - 1)
$varcomp          DF
[1]  0.05626259 70
$F.stats
   value     numdf     dendf    p.value
2176.141    14.000    70.000      0.000
$Rsquared
          R.squared    Adjusted.R.Squared
          0.9977076             0.9972491
$parameter
              Estimate Std. Error      t value       Pr(>|t|)
Intercept   3.45668380 0.02588035  133.5640100   5.003902e-86
Age 20     -5.75926889 0.09331294  -61.7199371   8.985629e-63
Age 25     -4.93096668 0.09331294  -52.8433311   3.765665e-58
Age 30     -3.55030012 0.09331294  -38.0472425   1.681907e-48
Age 35     -2.22896097 0.09331294  -23.8869436   2.241044e-35
Age 40     -1.01441116 0.09331294  -10.8710662   1.105870e-16
Age 45     -0.03049295 0.09331294   -0.3267816   7.448082e-01
Age 50      0.77114499 0.09331294    8.2640732   5.960482e-12
Age 55      1.42607568 0.09331294   15.2827213   5.404638e-24
Age 60      1.95382170 0.09331294   20.9383786   7.409972e-32
Age 65      2.35347336 0.09331294   25.2212964   7.373366e-37
Age 70      2.63356367 0.09331294   28.2229198   5.643536e-40
Age 75      2.79162262 0.09331294   29.9167787   1.298619e-41
Age 80      2.85054598 0.09331294   30.5482383   3.334580e-42
Age 85      2.73415278 0.09331294   29.3008958   0.000000e+00
```

TABLE 3.2
Output of Linear Period Effect Model by **R** Function `apclinfit` on Lung
Cancer Mortality among US Males

```
> apclinfit(r=y, apcmodel="P", header=T, transform="log",
    Plot=TRUE)
$model
lm(formula = log(rr) ~ x - 1)
$varcomp       DF
[1]   9.129288 78
$F.stats
       value          numdf          dendf        p.value
1.836485e+01 6.000000e+00 7.800000e+01 7.043255e-13
$Rsquared
            R.squared    Adjusted.R.Squared
            0.5855233              0.5536405
$parameter
               Estimate Std. Error    t value      Pr(>|t|)
Intercept     3.45668380  0.3296695 10.48529933 1.485812e-16
Period 1980   0.17666000  0.7371635  0.23964832 8.112315e-01
Period 1985   0.16418675  0.7371635  0.22272773 8.243296e-01
Period 1990   0.09045242  0.7371635  0.12270333 9.026576e-01
Period 1995  -0.01579687  0.7371635 -0.02142926 9.829580e-01
Period 2000  -0.14238876  0.7371635 -0.19315763 8.473376e-01
Period 2005  -0.27311354  0.7371635 -0.37049249 7.120198e-01
```

TABLE 3.3
Comparison of Linear Single Factor and Two Factor Models on Lung Cancer
Mortality Rate among US Males

Model	R.squared	Adjusted.R.Squared	Nparameters
A	0.9977076	0.9972491	14
P	0.5855233	0.5536405	6
C	0.9401313	0.9226312	19
AP	0.9990234	0.9987379	19
AC	0.9998313	0.9997275	32

TABLE 3.4
Output of Linear Age-Period Model by **R** Function `apclinfit` on Lung Cancer Mortality

```
> apclinfit(r=y, apcmodel="AP", header=T, transform="log",
    Plot=TRUE)
$model
lm(formula = log(rr) ~ x - 1)
$varcomp          DF
[1]  0.02581388 65
$F.stats
   value     numdf     dendf   p.value
3499.453    19.000    65.000     0.000
$Rsquared
           R.squared     Adjusted.R.Squared
           0.9990234              0.9987379
$parameter
              Estimate Std. Error    t value     Pr(>|t|)
Intercept    3.45668380 0.01753020 197.1844542 5.291139e-92
Age 20      -5.75926889 0.06320605 -91.1189482 2.694663e-70
Age 25      -4.93096668 0.06320605 -78.0141552 5.950048e-66
Age 30      -3.55030012 0.06320605 -56.1702569 8.166717e-57
Age 35      -2.22896097 0.06320605 -35.2649934 4.244085e-44
Age 40      -1.01441116 0.06320605 -16.0492729 2.691900e-24
Age 45      -0.03049295 0.06320605  -0.4824372 6.311161e-01
Age 50       0.77114499 0.06320605  12.2004930 1.822655e-18
Age 55       1.42607568 0.06320605  22.5623283 1.873651e-32
Age 60       1.95382170 0.06320605  30.9119407 1.381211e-40
Age 65       2.35347336 0.06320605  37.2349374 1.461056e-45
Age 70       2.63356367 0.06320605  41.6663219 1.307360e-48
Age 75       2.79162262 0.06320605  44.1670153 3.360321e-50
Age 80       2.85054598 0.06320605  45.0992576 9.017788e-51
Age 85       2.73415278 0.06320605  43.2577694 0.000000e+00
Period 1980  0.17666000 0.03919873   4.5067787 2.809958e-05
Period 1985  0.16418675 0.03919873   4.1885733 8.630708e-05
Period 1990  0.09045242 0.03919873   2.3075345 2.421894e-02
Period 1995 -0.01579687 0.03919873  -0.4029944 6.882748e-01
Period 2000 -0.14238876 0.03919873  -3.6324841 5.545939e-04
Period 2005 -0.27311354 0.03919873  -6.9674080 1.980533e-09
```

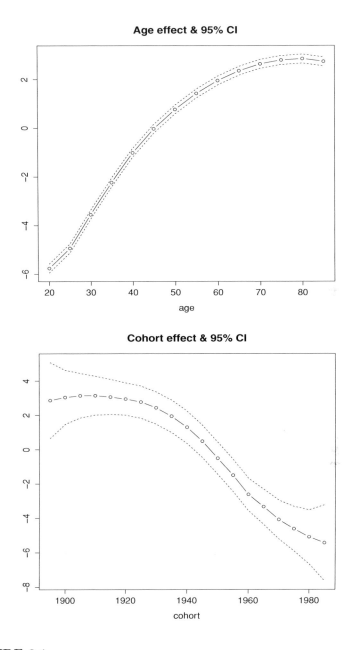

FIGURE 3.1

Plot by single factor linear model on lung cancer mortality rate among US males. Upper panel: age effect model; Lower panel: cohort effect model.

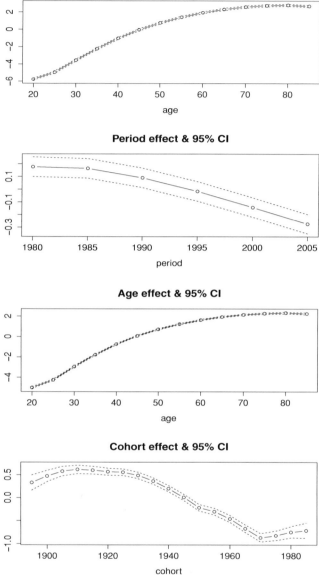

FIGURE 3.2
Plot by two factor linear models on lung cancer mortality rates among US males. Upper panels: age-period model; Lower panels: age-cohort model.

It is interesting to observe with the model comparison in Table 3.3 that although the age effect model has fewer parameters than the cohort effect model, it fits the data better in terms of the model goodness-of-fit with greater R^2 (0.9977) and adjusted R_a^2 (0.9972) than the cohort effect model with R^2 (0.9401) and adjusted R_a^2 (0.9226), respectively. Also notice the much lower R^2 and adjusted R_a^2 (< 0.6) by the period effect model though its F-test shows significance of the period effect in Table 3.2. This indicates that age effect is the most significant among all three factors of the age, period, and cohort in the study. Furthermore, both two factor models have greater adjusted R_a^2 than the age effect model, indicating the need to consider two factor models. It is also observed that the age-cohort (AC) model fits the data well with both R^2 and adjusted $R_a^2 > 0.9997$.

It is worthwhile to point out that the standard errors of age effects remain constant in the age effect model (Table 3.1). Similarly, the standard errors of period effects also remain constant in the period effect model (Table 3.2). Furthermore, the standard errors of age effects and period effects in the age-period model remain constant as well (Table 3.4). However, the standard errors of cohort effects vary in the cohort effect model, but display some symmetry from old cohorts to young cohorts in Figure 3.1. Such an observation of constant standard errors in the age effect model, period effect model, and age-period model reflects balanced data structure in the age and period. The varying standard errors of cohort effects reflect unbalanced but symmetric data structure in the cohorts. Recall that the ANOVA model has a design matrix with entries of $1, 0$, and -1 only. By linear model theory, the variance-covariance matrix (thus the standard error) of parameter estimates is completely determined by the design matrix and the variance component.

TABLE 3.5

Output of Linear Age-Cohort Effect Model by **R** Function `apclinfit` on Lung Cancer Mortality

```
> apclinfit(r=y, apcmodel="AC", header=T, transform="log",
    Plot=TRUE)
$varcomp            DF
[1]   0.005572459 52
$F.stats
    value    numdf    dendf   p.value
 9632.995   32.000   52.000     0.000
$Rsquared
            R.squared    Adjusted.R.Squared
            0.9998313             0.9997275
```

$parameter	Estimate	Std. Error	t value	Pr(>|t|)
Intercept	3.405558076	0.01055132	322.761214	1.550876e-87
Age 20	-4.994846464	0.04583050	-108.985209	4.638862e-63
Age 25	-4.235340818	0.04181760	-101.281304	2.062416e-61
Age 30	-2.942953922	0.03885795	-75.736215	6.825477e-55
Age 35	-1.760216030	0.03638742	-48.374307	6.522192e-45
Age 40	-0.723848761	0.03438077	-21.053886	4.123651e-27
Age 45	0.087096492	0.03295705	2.642727	1.084140e-02
Age 50	0.730841590	0.03228374	22.638070	1.365224e-28
Age 55	1.242756054	0.03228374	38.494797	6.835878e-40
Age 60	1.639465266	0.03295705	49.745504	1.571332e-45
Age 65	1.940450927	0.03438077	56.440012	2.488603e-48
Age 70	2.151276034	0.03638742	59.121429	2.309621e-49
Age 75	2.273872416	0.03885795	58.517562	3.908617e-49
Age 80	2.335106281	0.04181760	55.840278	4.299000e-48
Age 85	2.256340934	0.04583050	49.232302	0.000000e+00
Cohort 1895	0.331813010	0.08365410	3.966489	2.241994e-04
Cohort 1900	0.474647867	0.06218977	7.632250	4.870699e-10
Cohort 1905	0.579046343	0.05247634	11.034427	3.113055e-15
Cohort 1910	0.617219047	0.04635666	13.314572	2.198105e-18
Cohort 1915	0.601476288	0.04184300	14.374598	9.442094e-20
Cohort 1920	0.569422874	0.03819844	14.906965	2.047645e-20
Cohort 1925	0.557580095	0.03679092	15.155371	1.015271e-20
Cohort 1930	0.488510932	0.03566480	13.697287	6.941654e-19
Cohort 1935	0.366270930	0.03493302	10.484950	1.974745e-14
Cohort 1940	0.201627805	0.03467921	5.814083	3.779625e-07
Cohort 1945	0.009480331	0.03493302	0.271386	7.871687e-01

```
$parameter        Estimate Std. Error      t value      Pr(>|t|)
Cohort 1950 -0.216798011 0.03566480   -6.078768 1.445571e-07
Cohort 1955 -0.300517280 0.03679092   -8.168245 6.908175e-11
Cohort 1960 -0.458846076 0.03819844  -12.012166 1.277728e-16
Cohort 1965 -0.671566841 0.04184300  -16.049682 8.624089e-22
Cohort 1970 -0.867467453 0.04635666  -18.712899 9.377595e-25
Cohort 1975 -0.822127218 0.05247634  -15.666626 2.451726e-21
Cohort 1980 -0.746475938 0.06218977  -12.003195 1.314985e-16
Cohort 1985 -0.713296705 0.08365410   -8.526740 1.890221e-11
```

Hence, balanced data structure results in balanced variance, leading to either constant standard errors or symmetric but varying standard errors.

Readers are also advised that such symmetric pattern in standard errors is true only for linear models. For generalized linear models, such as the loglinear model, the symmetric pattern in standard errors may not be true since the population exposure may have unbalanced structure. I will demonstrate in the next section that the constant or symmetric standard errors do not hold for loglinear models because of unbalanced population structure.

3.2 Loglinear Models for Discrete Response

I now consider modeling the event data with more than one piece of information in each cell, often two out of three pieces of the event rate, event frequency count, and population exposure. For example, Tables 1.1 and 1.2 display both the number of deaths and the event rate in each cell, where the latter is calculated with the event frequency count divided by the population exposure in the cell. When two of the three pieces are available, modeling the event rate without incorporating the information of the number of cases or the population exposure is incomplete because the extra piece of information, if properly incorporated in the modeling, may often improve the accuracy and efficiency of parameter estimation.

A loglinear model can serve the needs by incorporating the extra piece of information through a Poisson distribution. It assumes that the number of cases out of the known population exposure in each cell follows a Poisson distribution, whose log-intensity is a linear function of age, period, and cohort effects. A loglinear model with either a single factor or two factors can be fitted to estimate the age, period, and cohort effects. Since the model fitting is implemented through the maximum likelihood estimation (MLE) approach for generalized linear models, the parameter estimates are slightly biased for finite samples, but consistent following linear model theory (McCullagh and Nelder 1989). This means that the parameter estimates by the loglinear model converge in probability to true parameter values as the sample size diverges to infinity. Furthermore, the estimates follow an approximately normal distribution with the true parameters as the mean. Interested readers may find the consistency results for generalized linear models in the monograph by McCullagh and Nelder (1989).

However, the consistency in APC models may not be concluded directly based on linear model theory because the number of parameters in the single factor cohort effect, two factor, or three factor models diverges to infinity as the sample size goes to infinity. Notice that APC data has virtually one data point in each cell. Although three pieces of information (the rate, frequency count, and population exposure) are available in a cell, the divergence of sample size in the cell (the population exposure) only ensures high accuracy of the rate, but has no impact on the estimates of age, period, and cohort effects. Thus, the consistency of parameter estimates requires that the sample size go to infinity, which is achieved only as the number of age groups or periods goes to infinity, or both. Since human life span is limited, it is preferred to let the number of periods (p) go to infinity, consequently, the sample size (i.e. the number of subjects involved in the study) diverges to infinity by incorporating more populations across a large number of periods in the study. This further raises questions about the variance estimation of the effect estimates in APC loglinear models. I will address these issues with the large sample asymptotic studies in Chapters 8 and 9. I now introduce loglinear models with a single factor or two factors.

3.2.1   Single Factor Models

Age effect model The age effect model assumes that the number of cases out of the population exposure N_{ij} in cell (i, j) follows a Poisson distribution whose log-intensity is a linear function of the age effect α_i

$$\log(E_{ij}) = \log(N_{ij}) + \mu + \alpha_i , \tag{3.8}$$

where E_{ij} is the expected event frequency in cell (i, j), and $\log(N_{ij})$ is the model offset term by the population exposure. The parameters μ and $\alpha_i, i = 1, \ldots, a$, are the overall mean effect and the i-th age group effect, respectively. For parameter estimation, I require the centralization $\sum_{i=1}^{a} \alpha_i = 0$ for efficient estimation, as discussed in the last section. The data structure is balanced with the same number (p) of observations for each age effect. The model yields consistent estimates as the sample size diverges to infinity, specifically, as the number of observations p in each age group tends to infinity.

The model goodness-of-fit is often assessed through a comparison of the model deviance with the degrees of freedom (df) of the loglinear model. The model has a parameters and $a(p-1)$ df for the deviance. A model deviance close to the df $a(p-1)$ indicates a good fit, while a large deviance relative to the df may indicate poor model fitting.

Period effect model A period effect model assumes that the number of cases out of the population exposure N_{ij} in cell (i,j) follows a Poisson distribution whose log-intensity is a linear function of the period effect

$$\log(E_{ij}) = \log(N_{ij}) + \mu + \beta_j , \qquad (3.9)$$

where the notations E_{ij}, N_{ij} and μ remain the same as before. β_j is the j-th period effect and satisfies a centralization $\sum_{j=1}^{p} \beta_j = 0$. The data structure is balanced for the period effect model and each period effect has the same number (a) of observations. Similar to the age effect model, the parameter estimates obtained through the MLE approach are slightly biased for finite samples but consistent. The consistency of period effect estimates is concluded directly by linear model theory (McCullagh and Nelder 1989), similar to the age effect model (3.8), provided that one is willing to assume in theory, for the convenience of asymptotic studies, that the number of age groups diverges to infinity. The model has p parameters and thus $p(a-1)$ degrees of freedom for the deviance. A deviance close to the df $p(a-1)$ indicates a good fit to the data.

Cohort effect model Alternatively, one may also fit a single factor cohort effect model (3.10), assuming that the number of cases out of the population exposure N_{ij} in cell (i,j) follows a Poisson distribution, whose log-intensity is a linear function of the cohort effect

$$\log(E_{ij}) = \log(N_{ij}) + \mu + \gamma_k , \qquad (3.10)$$

where the notations E_{ij}, N_{ij} and μ remain the same as before. γ_k is the k-th cohort effect and satisfies the centralization $\sum_{k=1}^{a+p-1} \gamma_k = 0$ with $a-i+j = k$ for each fixed k. The data structure is unbalanced for the cohort effect model and each cohort parameter γ_k has a varying number (m_k) of observations, from 1 on the oldest and youngest cohorts to a maximum, $\min(a,p)$, on the central diagonals. The cohort effect model has $a + p - 1$ parameters and thus has $(a - 1)(p - 1)$ degrees of freedom for the residual deviance.

The cohort effect model has a parameter estimation issue since the consistency of parameter estimates is not supported. This is because the asymptotic study requires either $a \to \infty$ or $p \to \infty$, which leads to a diverging number of cohort effects and makes it difficult to establish the consistency results for non-Gaussian response variables. In practice, a two-factor model, the age-period model, is preferred. It has the same number of parameters and valid consistency of parameter estimation. Furthermore, it often performs better than the cohort effect model. Hence, the loglinear model with single factor cohort effect is not recommended in practice.

3.2.2 Two Factor Models

In addition to the above single factor models, if both age and period effects, or age and cohort effects, are significant, one may fit a two factor AP or AC model. Again for the same reason, I do not consider a period-cohort model because the age effects have been found to be significant. Readers may also explore in certain studies where age effects become relatively less important with no significance.

Age-period model An AP model assumes that the number of cases out of the population exposure N_{ij} in cell (i, j) follows a Poisson distribution whose log-intensity is a linear function of both age and period effects.

$$\log(E_{ij}) = \log(N_{ij}) + \mu + \alpha_i + \beta_j , \qquad (3.11)$$

where E_{ij}, N_{ij}, model parameters μ, α_i and β_j remain the same as before. For parameter estimation, a centralization is required for age and period effects $\sum_{i=1}^{a} \alpha_i = 0$ and $\sum_{j=1}^{p} \beta_j = 0$, respectively. The model is fitted through a maximum likelihood estimation approach and the MLE of the intercept, age and period effects are slightly biased for finite samples but consistent for diverging samples. The consistency is in the sense of the following setting. The intercept and age effect estimates converge to true parameter values as the number of periods (p) diverges to infinity. By symmetry, the intercept and period effect estimates converge to true parameter values as the number of age groups (a) diverges to infinity. The AP model has a balanced data structure, i.e. each age effect has p observations and each period effect has a observations. The model has $a + p - 1$ parameters and thus $(a-1)(p-1)$ degrees of freedom for the deviance. A model deviance close to the df indicates a good fit, while a relative large deviance to the df may indicate poor model fitting.

As discussed in the last subsection, the AP model has the same number of parameters and degrees of freedom as the cohort effect model, and often performs better than the cohort effect model. This can be explained as follows. For a response variable following an exponential family distribution (MuCullagh and Nedler 1989), such as the Poisson distribution in loglinear models, the consistency of the intercept, age, and period effect estimates can be achieved by large sample theory. In contrast, the cohort effect model (3.10) has no guaranteed consistency because of its diverging number of parameters, and may perform poorly.

Since the AP model achieves the consistency of the intercept and age effect estimates through the maximum profile likelihood with $p \to \infty$, the variance of the intercept and age effects can be calculated with asymptotic normality. The variance of the period effects can be calculated through the profile function with the Delta method; see Chapter 9. Although, the age and period are perpendicular on an APC rectangular table, the variance of period effect estimates depends on the age effects for non-Gaussian response variable. See Chapter 9 for more details.

Age-cohort model Alternatively, an AC model assumes that the number of cases out of the population exposure N_{ij} in cell (i,j) follows a Poisson distribution whose log-intensity is a linear function of both age and cohort effects.

$$\log(E_{ij}) = \log(N_{ij}) + \mu + \alpha_i + \gamma_k, \qquad (3.12)$$

where E_{ij}, N_{ij} and model parameters μ, α_i and γ_k remain the same as before. For parameter estimation, a centralization is required for age and cohort effects $\sum_{i=1}^{a} \alpha_i = 0$ and $\sum_{k=1}^{a+p-1} \gamma_k = 0$, respectively. Similar to the AP model, the AC model is fitted through a maximum profile likelihood estimation (MaPLE) approach, which yields consistent estimates for the intercept and age effects as the number of periods (p) diverges to infinity. The data structure is unbalanced, i.e. the cohort effect has a varying number (m_k) of observations, similar to the cohort effect model. Meanwhile the number of observations for each age effect remains the same (p). The model has $2a + p - 2$ parameters and $(a-1)(p-2)$ degrees of freedom for model deviance. A deviance close to the df $(a-1)(p-2)$ indicates a good fit while a relatively large deviance to the df may indicate poor model fitting.

The intercept and age effect estimates follow an asymptotically normal distribution through the MaPLE approach. The consistency of the MaPLE estimates of loglinear AC model (3.12) is a special case of loglinear APC model, which will be studied in Chapter 8. Interested readers are referred to the consistency of the MaPLE estimates by the large sample theory there. The variance of the cohort effects, however, needs to be estimated through a profile function with the Delta method because the variance may not be directly derived from the consistency of the MaPLE estimates. See more details of the Delta method in Chapter 9.

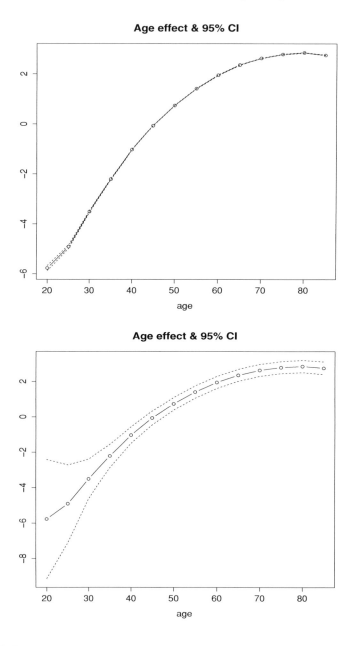

FIGURE 3.3
Plot of age effects and point-wise 95% confidence intervals by the loglinear model on lung cancer mortality among US males. Upper panel: maximum likelihood estimation; Lower panel: quali-likelihood estimation adjusting for over-dispersion.

3.2.3 Modeling Over-Dispersion with Quasi-Likelihood

The Poisson distribution requires that the expected value and the variance be identical. However, this may not be true for given count data. In practice, it is often observed that the variance presented in the count data is larger than the expected value. Such a phenomenon is called over-dispersion. When over-dispersion is present, fitting a loglinear model through the maximum likelihood (ML) approach is inappropriate because the variance is estimated with the expected value, resulting in a smaller variance than the true value. To correct the bias in the variance estimation, a quasi-likelihood (Q-L) approach fits the model with the same log link function, but specifies the variance as a multiple of the expected value and further estimates the multiplication factor. Often the multiplication factor is larger than 1, indicating over-dispersion. Hence, the Q-L approach yields proper standard errors of the parameter estimates, larger than the ones by the ML approach, when over-dispersion is present.

For the lung cancer mortality data among US males, over-dispersion is observed through the comparison of the age trend and 95% confidence intervals in Figure 3.3, in which the ML approach presents extremely narrow confidence intervals while the Q-L presents reasonably moderate confidence intervals, indicating strong over-dispersion in the data. Tables 3.6 and 3.7 display the parameter estimates and standard errors of the loglinear age effect model fitted with the ML and the Q-L approaches, respectively. The Q-L standard errors are much larger than the ML ones, indicating over-dispersion. Similarly, Table 3.9 illustrates the difference in standard errors of the period effect model, with moderate standard errors by the Q-L approach and smaller ones by the ML approach. Furthermore, Table 3.10 presents the output of the AP model fitted with the **R** function `apcglmfit` through the Q-L approach, including

TABLE 3.6
Output of Loglinear Age Effect Model on Lung Cancer Mortality Data among
US Males

```
apcglmfit(r=USlungCAmale.r, header=T, n.risk= USlungCAmale.p,
    Scale=1e-5, apcmodel="A", fam="loglin", Plot=T)
```

```
$model
glm(formula = rr ~ x + offset(log(n.risk)),
    family = poisson(link = log))
$deviance
     Deviance DF
[1,] 73061.63 70
$pearson.chisq
[1] 72377.06
$p.val
p.value
      0
$parameter
              Estimate  Std. Error      z value     Pr(>|z|)
Intercept -8.05564287 0.005145936 -1565.43791 0.000000e+00
Age 20    -5.76489162 0.053348191  -108.06161 0.000000e+00
Age 25    -4.90562899 0.034884908  -140.62324 0.000000e+00
Age 30    -3.50675259 0.017847197  -196.48758 0.000000e+00
Age 35    -2.20570775 0.010410497  -211.87343 0.000000e+00
Age 40    -1.02687782 0.007317302  -140.33559 0.000000e+00
Age 45    -0.06588452 0.006168916   -10.68008 1.261618e-26
Age 50     0.73884890 0.005690491   129.83922 0.000000e+00
Age 55     1.40748424 0.005475611   257.04608 0.000000e+00
Age 60     1.94959765 0.005372941   362.85482 0.000000e+00
Age 65     2.35407651 0.005326897   441.92270 0.000000e+00
Age 70     2.63169868 0.005317719   494.89243 0.000000e+00
Age 75     2.78775826 0.005345486   521.51630 0.000000e+00
Age 80     2.85227456 0.005449762   523.37595 0.000000e+00
Age 85     2.75400448 0.005642063   488.12008 0.000000e+00
```

model deviance, degrees of freedom (df), and the number of parameters of the
model. Notice that unlike the linear model shown in Table 3.4, the loglinear
model does not yield constant standard errors, but rather varying standard
errors. Figure 3.4 presents trend estimation by the AP and AC models, where
the large confidence intervals in the early age groups and young cohorts may

TABLE 3.7

Output of Loglinear Age Effect Model with Quasi-Likelihood on Lung Cancer
Mortality Data among US Males

```
apcglmfit(r=USlungCAmale.r, header=T, n.risk= USlungCAmale.p,
    Scale=1e-5, apcmodel="A", fam="qlik", Plot=T)

$model
glm(formula = rr ~ x + offset(log(n.risk)),
    family = quasipoisson(link = log))
$deviance
     Deviance DF
[1,] 73061.63 70
$pearson.chisq
[1] 72377.06
$p.val
p.value
      0
$dispersion
[1] 1033.958
$parameter
```

	Estimate	Std. Error	t value	Pr(>|t|)
Intercept	-8.05564287	0.1654687	-48.6837881	1.007111e-55
Age 20	-5.76489162	1.7154228	-3.3606243	1.262458e-03
Age 25	-4.90562899	1.1217319	-4.3732634	4.174058e-05
Age 30	-3.50675259	0.5738805	-6.1105968	4.992614e-08
Age 35	-2.20570775	0.3347518	-6.5890837	6.949165e-09
Age 40	-1.02687782	0.2352894	-4.3643174	4.310429e-05
Age 45	-0.06588452	0.1983629	-0.3321414	7.407750e-01
Age 50	0.73884890	0.1829790	4.0378894	1.361555e-04
Age 55	1.40748424	0.1760695	7.9939145	1.874578e-11
Age 60	1.94959765	0.1727681	11.2844764	2.068505e-17
Age 65	2.35407651	0.1712875	13.7434203	1.458738e-21
Age 70	2.63169868	0.1709924	15.3907338	3.690542e-24
Age 75	2.78775826	0.1718853	16.2187137	2.080837e-25
Age 80	2.85227456	0.1752383	16.2765471	1.707632e-25
Age 85	2.75400448	0.1814218	15.1801196	0.000000e+00

be explained by the low mortality rate and the relatively small numbers of
deaths in the early age groups and young cohorts, which presents relatively
large variability in estimation.

TABLE 3.8

Comparison of Loglinear Single Factor and Two Factor Models on Lung Cancer Mortality

Model	Deviance	DF	Nparameters
A	73061.63	70	14
P	5649156	78	6
C	664847.2	65	19
AP	28506.15	65	19
AC	1471.423	52	32

Table 3.11 displays the output of the **R** function `apcglmfit` on the lung cancer mortality data among US males with a loglinear AC model through a MaPLE approach with quasi-likelihood for the standard errors of the intercept and age effects, while using the Delta method for the cohort effect estimates. The over-dispersion is severe with a dispersion parameter 28.27, which yields fairly wide confidence intervals for age and cohort estimates.

Table 3.8 displays model deviance, degrees of freedom for single factor, and two factor loglinear models for comparison. It is shown that the period effect model (P) has the largest deviance (5649156) on 78 degrees of freedom (df), followed by the cohort effect model (C) with deviance (664847.2) on 65 df and the age effect model (A) with deviance (73061.63) on 70 df. The AP and AC models have better goodness-of-fit with much smaller deviance 28506 and 1471.42 on 65 and 52 df, respectively. It is also shown that the cohort effect model and the AP model have the same number of parameters (thus the same df), but their goodness-of-fit differs largely, indicating different model performance. Lastly, the large difference in the model deviance between the AP and AC models also indicates a major difference between the period and cohort effects in the presence of the age effect. This may further imply that a full APC model including all three factors is needed to achieve the best fit.

3.2.4   R Programming for Loglinear Models

I now provide **R** program on the loglinear model with the function `apcglmfit`.

R program for loglinear model on frequency, population exposure

Fit a single factor model

Make header for rate data. Population data needs no header.

```
x=apcheader(r=USlungCAmale.r, agestart=20, yearstart=1960,
 agespan=5, yearspan=5, head=F)
```

Fit age effect model and plot the age trend.

```
apcglmfit(r=y, header=T, n.risk=USlungCAmale.p, apcmodel="A",
 fam="loglin", Plot=T)
```

Fit age effect model with quasi-likelihood and plot the trend.

```
apcglmfit(r=y, header=T, n.risk=USlungCAmale.p, apcmodel="A",
 fam="qlik", Plot=T)
```

Fit period effect model with quasi-likelihood and plot the trend.

```
apcglmfit(r=y, header=T, n.risk=USlungCAmale.p, apcmodel="P",
 fam="qlik", Plot=T)
```

Fit cohort effect model with quasi-likelihood and plot the trend.

```
apcglmfit(r=y, header=T, n.risk=USlungCAmale.p, apcmodel="C",
 fam="qlik", Plot=T)
```

The option `fam` specifies the method of model fitting, either `loglin` as the default for maximum likelihood or `qlik` for quasi-likelihood.

Fit a two factor model

```
apcglmfit(r=y, header=T, n.risk=USlungCAmale.p, apcmodel="AP",
 fam="loglin", Plot=T) # Fit AP model with ML and plot the trends.
apcglmfit(r=y, header=T, n.risk=USlungCAmale.p, apcmodel="AC",
 fam="qlik", Plot=T) # Fit AC model with Q-L and plot the trends.
```

TABLE 3.9

Output of Loglinear Period Effect Model on Lung Cancer Mortality Data among US Males

```
########## Loglinear Model --- ML Fit ###########
$model
glm(formula = rr ~ x + offset(log(n.risk)),
    family = poisson(link = log))
$deviance
     Deviance DF
[1,]  5649156 78
$parameter
                Estimate    Std. Error       z value    Pr(>|z|)
Intercept    -6.934800563 0.0006169667 -11240.15339 0.00000000
Period 1980   0.060841930 0.0014401536     42.24683 0.00000000
Period 1985   0.083347453 0.0013860748     60.13200 0.00000000
Period 1990   0.073543814 0.0013549087     54.27953 0.00000000
Period 1995   0.002940646 0.0013582157      2.16508 0.03038154
Period 2000  -0.073204221 0.0013642355    -53.65952 0.00000000
Period 2005  -0.147469621 0.0013720519   -107.48108 0.00000000
```

```
########## Loglinear Model --- Q-L Fit ###########
$model
glm(formula = rr ~ x + offset(log(n.risk)),
    family = quasipoisson(link = log))
$deviance
     Deviance DF
[1,]  5649156 78
$dispersion
[1] 89635.67
$parameter
                Estimate Std. Error       t value     Pr(>|t|)
Intercept    -6.934800563  0.1847150 -37.543244777 1.076016e-51
Period 1980   0.060841930  0.4311707   0.141108685 8.881479e-01
Period 1985   0.083347453  0.4149799   0.200846951 8.413411e-01
Period 1990   0.073543814  0.4056491   0.181299108 8.566031e-01
Period 1995   0.002940646  0.4066391   0.007231586 9.942485e-01
Period 2000  -0.073204221  0.4084414  -0.179228199 8.582233e-01
Period 2005  -0.147469621  0.4107816  -0.358997640 7.205668e-01
```

TABLE 3.10

Output of Loglinear AP Model on Lung Cancer Mortality Data among US Males

```
$model
glm(formula = rr ~ x + offset(log(n.risk)),
    family = quasipoisson(link = log))
$deviance
    Deviance DF
[1,] 28506.15 65
$pearson.chisq
[1] 28650.73
$p.val
p.value
      0
$dispersion
[1] 440.7804
$parameter
```

	Estimate	Std. Error	t value	Pr(>|t|)
Intercept	-8.047282107	0.10803982	-74.4844110	1.167600e-64
Age 20	-5.779275243	1.12003485	-5.1599066	2.522541e-06
Age 25	-4.924542105	0.73240367	-6.7238086	5.320847e-09
Age 30	-3.525007306	0.37470203	-9.4074946	9.546142e-14
Age 35	-2.214417729	0.21856932	-10.1314205	5.287570e-15
Age 40	-1.024215849	0.15363033	-6.6667555	6.702063e-09
Age 45	-0.052228393	0.12952554	-0.4032285	6.881035e-01
Age 50	0.754788940	0.11948172	6.3171918	2.736799e-08
Age 55	1.418058475	0.11496754	12.3344243	1.106884e-18
Age 60	1.948912363	0.11280901	17.2762112	5.492249e-26
Age 65	2.345793586	0.11184148	20.9742711	1.223866e-30
Age 70	2.625224575	0.11164633	23.5137562	1.706505e-33
Age 75	2.789529610	0.11222891	24.8557137	6.604908e-35
Age 80	2.865318203	0.11442548	25.0409109	4.264169e-35
Age 85	2.772060873	0.11846738	23.3993603	0.000000e+00
Period 1980	0.101112999	0.03026455	3.3409715	1.386932e-03
Period 1985	0.123582505	0.02911925	4.2440136	7.117855e-05
Period 1990	0.101201417	0.02845872	3.5560776	7.081458e-04
Period 1995	0.005589874	0.02852911	0.1959358	8.452717e-01
Period 2000	-0.100798074	0.02866503	-3.5164129	8.030246e-04
Period 2005	-0.230688720	0.02884084	-7.9986824	2.934408e-11

TABLE 3.11
Output of Loglinear AC Model on Lung Cancer Mortality Data among US
Males

```
apcglmfit(r=USlungCAmale.r, header=T, n.risk= USlungCAmale.p,
    Scale=1e-5, apcmodel="AC", fam="qlik", Plot=T)
$model
glm(formula = rr ~ x - 1 + offset(log(n.risk)),
    family = quasipoisson(link = log))
$deviance
     Deviance DF
[1,] 1471.423 52
$pearson.chisq
[1] 1470.237
$p.val
p.value
      0
$dispersion
[1] 28.27379
```

	Estimate	Std. Error	t value	Pr(>|t|)
Intercept	-8.11681800	0.04744114	-171.0923729	3.237979e-73
Age 20	-4.96937518	0.32716687	-15.1891151	9.235085e-21
Age 25	-4.19921432	0.19826445	-21.1798647	3.121513e-27
Age 30	-2.90229907	0.10013781	-28.9830480	9.136529e-34
Age 35	-1.74039683	0.05958085	-29.2106758	6.225057e-34
Age 40	-0.72709318	0.04371012	-16.6344369	1.806703e-22
Age 45	0.07466343	0.03828181	1.9503633	5.653254e-02
Age 50	0.71981792	0.03634893	19.8029999	7.045637e-26
Age 55	1.22862882	0.03560122	34.5108641	1.624126e-37
Age 60	1.62722016	0.03532132	46.0690621	7.805254e-44
Age 65	1.92770038	0.03527712	54.6444921	1.300733e-47
Age 70	2.13978537	0.03538143	60.4776452	7.220217e-50
Age 75	2.26054683	0.03565476	63.4009902	6.404638e-51
Age 80	2.31855921	0.03630891	63.8564721	4.434503e-51
Age 85	2.24145645	0.03755730	59.6809799	0.000000e+00
Cohort 1895	0.35617430	0.05737630	6.2076899	9.033755e-08
Cohort 1900	0.49941641	0.05751161	8.6837494	1.074918e-11
Cohort 1905	0.59840678	0.05749709	10.4076018	2.575717e-14
Cohort 1910	0.63559377	0.05745081	11.0632692	2.664535e-15
Cohort 1915	0.62550622	0.05739810	10.8976820	4.884981e-15

	Estimate	Std. Error	t value	Pr(>|t|)
Cohort 1920	0.59523119	0.05734759	10.3793576	2.842171e-14
Cohort 1925	0.58553466	0.05730407	10.2180298	4.907186e-14
Cohort 1930	0.51136438	0.05723491	8.9344845	4.383605e-12
Cohort 1935	0.38240890	0.05714070	6.6924089	1.530979e-08
Cohort 1940	0.22332964	0.05702839	3.9161135	2.634870e-04
Cohort 1945	0.02419744	0.05689298	0.4253149	6.723621e-01
Cohort 1950	-0.18917891	0.05669157	-3.3369849	1.569619e-03
Cohort 1955	-0.28553180	0.05638397	-5.0640597	5.502255e-06
Cohort 1960	-0.42154093	0.05551978	-7.5926263	5.630478e-10
Cohort 1965	-0.70035081	0.05388348	-12.9975048	0.000000e+00
Cohort 1970	-0.99040125	0.05120237	-19.3428790	0.000000e+00
Cohort 1975	-0.93732084	0.08373989	-11.1932421	1.776357e-15
Cohort 1980	-0.78152209	0.23135372	-3.3780399	1.389622e-03
Cohort 1985	-0.73131706	0.62841265	-1.1637529	2.498362e-01

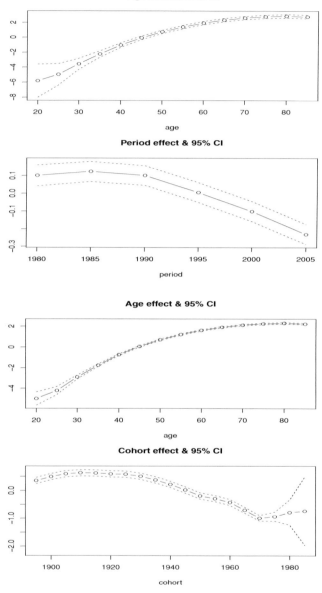

FIGURE 3.4
Comparison of effect estimates and 95% confidence intervals by loglinear model with quasi-likelihood approach on lung cancer mortality among US males. Upper panels: AP model; Lower panels: AC model.

3.3 Suggested Readings

Early work on the single factor models and two factor models may be found in a number of review articles, including Kupper et al (1983, 1985), Janis et al (1985), Clayton and Schifflers (1987), Fienberg and Mason (1985), Holford (1985, 1991), Robertson and Boyle (1998b), and Rodgers (1982). Examination of model diagnostics through the residual plot is highly recommended for fitting linear models (Sen and Srivastava 1990). Generalized linear models and over-dispersion issues are well explained in McCullagh and Nelder (1989). The profile likelihood approach to APC models can be found in Fu (2016).

3.4 Exercises

1. Fit a linear cohort effect model to the US male lung cancer mortality rate data. Examine the standard error of the estimates for the cohort effects. Compare the model goodness-of-fit with the age effect model in Table (3.1), the period effect model in Table (3.2) and the age-period model in Table (3.4).

2. Fit the following linear model to the US female lung cancer mortality rate data, including the age effect (A) model, the period effect (P) model, the cohort effect (C) model, the AP model, the AC model. Compare the goodness-of-fit with the corresponding model to the US male mortality rate data. Examine the standard error of the estimates for the age, the period and the cohort effects.

3. Fit a loglinear cohort effect model to the US male lung cancer mortality rate data with the quasi-likelihood approach. Examine the standard error of

the estimates for the cohort effects. Compare the model deviance with the loglinear age-period model with the quasi-likelihood approach in Table (3.10).

4. Fit a loglinear age-cohort model to the US male lung cancer mortality rate data with the quasi-likelihood approach. Examine the standard error of the estimates for age and cohort effects. Compare the model deviance and degrees of freedom with the age effect model in Table (3.7), the cohort effect model in the above exercise 3, and the age-period model in (3.10) with the quasi-likelihood approach.

5. Fit the following loglinear models to the US female lung cancer mortality data through the quasi-likelihood approach, including the age effect (A) model, the period effect (P) model, the cohort effect (C) model, the age-period (AP) model, and the age-cohort (AC) model. Examine the standard error of the estimates for the age, the period and the cohort effects. Compare the model deviance and degrees of freedom with the corresponding model for the US male lung cancer mortality data.

6.* Construct a design matrix X for the linear age-period model. Further show that the variance of the age effect remains the same across different levels, and the same holds true for period effects.

7.* Show that for fixed $i = 1, \ldots, a$ and $k = 1, \ldots, a + p - 1$, the least-squares estimates for the parameters of the AC model satisfy the following equations.

$$\mu + \alpha_i + \frac{1}{p} \sum_{k=a-i+1}^{a-i+p} \gamma_k = y_i.$$

and

$$\mu + \gamma_k + \frac{1}{m_k} \sum_{a-i+j=k} \alpha_i = \frac{1}{m_k} \sum_{a-i+j=k} y_{ij} \ .$$

---

* Difficult exercises with an asterisk are meant for graduate students in biostatistics.

8.* Conduct a simulation study on the consistency of the loglinear model with single factor cohort effect (3.10) by generating Poisson counts, and compare the results with those from two factor loglinear model with age-cohort effects (3.12). Explain the similarities and differences between the two models.

4

Age-Period-Cohort Models — Complexity with Linearly Dependent Covariates

I have shown so far that models with age, period, and cohort effects present interesting secular trends in the age, period, and birth cohort, with either single factor models or two factor models. For a given data set, models with different factors may present different trends. For example, in the loglinear AP or AC model for the US male lung cancer mortality data, the age effect estimate differs by the presence of the second factor, the period or the cohort (Figure 3.4). This is because the special pattern of the age, period, and cohort in the rectangular table determines inherent relationship among them. In this chapter, I study the pattern of the APC data in the rectangular tables and illustrate the relationship among the age, period, and cohort. I also show that such inherent relationship leads to model complexity and induces a complex parameter identification problem in age-period-cohort models.

4.1 Lexis Diagram and Patterns in Age, Period, and Cohort

In this section, I study special patterns in APC data, i.e. the so-called Lexis diagram, paying special attention to age, period, and cohort groups. I then study two different patterns, the regular one that has identical time span in both age groups and periods (e.g. the lung cancer data in Tables 1.1 and 1.2), and the irregular one that has different time spans in age groups from periods (e.g. the mean value of retirement accounts data in Table 1.4). The latter case requires special parameter estimation in age-period-cohort models.

4.1.1   Lexis Diagram and Dependence among Age, Period, and Cohort

A typical APC data set is often displayed in a rectangular table of a rows and p columns. A special pattern of the table was noticed and named the Lexis diagram after the late German statistician and sociological economist Wilhelm Lexis. In the Lexis diagram, a simultaneous move-up or move-down in the row and column by one cell will land on the same diagonal of the table. Such a special pattern determines a linear relationship among the rows (age), columns (period) and diagonals (cohort). It is often referred to in the APC literature as **Period $-$ Age $=$ Cohort**. However, such description is an approximation. In fact, each of the age, period, and cohort factors is categorized into a number of levels, and expressed with the fixed effects of these levels. A linear dependence exists among the indices (expressed with dummy variables) of the age, period, and cohort levels, but not of the factors themselves, since the indices appear in the APC models but not directly the factors themselves as a whole. It is well known that regression models with three linearly dependent continuous covariates need to drop one in order to resolve the identification problem. However, the APC models with discrete linearly dependent covariates can resolve the identification problem with all covariates involved in the model because of the increased parameter space and the model degrees of freedom through discretization of the factors.

4.1.2   Explicit Pattern in APC Data with Identical Spans in Age and Period

To make this special APC pattern and the Lexis diagram easy to understand, I shall take an example of the US male lung cancer mortality data. As shown in Table 1.1, the table has 14 rows of age groups from 20–24 to 80–84 and 85+, 6 columns of period groups from 1980–84 to 2005–09, and 19 diagonals of birth cohorts with mid birth year 1895 of the oldest generation to 1985 of the youngest. The first cell in the table has a mortality rate of 0.1 per 100,000 person-year with 68 deaths. It means that 68 males died from lung cancer among those US men of age 20–24 during 1980–84. These men were born around the year 1960, more precisely between 1956 and 1964. Now consider lung cancer death among men aged 25–29 during the years 1985–89; these men were also born between 1956 and 1964. It indicates that a simultaneous shift up by one level (5 years in each level) in the age group and period makes the diagonal (birth cohort) unchanged.

The above pattern is explicit since birth cohorts follow the same pattern as the diagonals of the table, as shown in Table 4.1 of dimension 3×3 with equal spans in age groups and periods. This is simply because the age and period have identical time spans in each group, which is 5 years in both. However, when age groups and periods have unequal time spans, the cohort pattern becomes complex, which will be discussed in the next subsection.

4.1.3   Implicit Pattern in APC Data with Unequal Spans in Age and Period

The APC data may not always have identical time spans in age groups and periods. A good example is the social survey data of US family mean value of retirement accounts, which has 10 years in each age group and 3 years apart between periods. The diagonals of the table do not define birth cohorts,

TABLE 4.1

Illustration of the Cohort Pattern in APC Tables of Equal Spans in Age and Period*

	Period		
Age	1980–1984	1985–1989	1990–1994
20–24	coh 1960	coh 1965	coh 1970
25–29	coh 1955	coh 1960	coh 1965
30–34	coh 1950	coh 1955	coh 1960

* Mid-birth year of cohorts is shown in cells.

since a cohort requires matching time spans in both age and period, requiring that multiple periods be matched to each age group. In order to estimate birth cohort effects, one needs to model the age, period, and cohort effects by coding birth cohort levels in the data through a special coding scheme. For example, in the social survey data with 10 year age span and 3 years apart between periods, one would expect to match each 10 year age span with 10 years in period by pooling 3 periods together. Ideally, 3 periods need to be collapsed into one to form a table with identical spans. For the current table with unequal spans, it requires coding 3 consecutive periods with one cohort level within one age group. Apparently, such a pattern in age, period, and cohort does not follow the pattern of rows, columns, and diagonals of the table, and is thus an implicit pattern.

Another case of unequal age and period spans is even more complex for coding cohort effects. It occurs when neither age span nor period span is roughly a multiple of the other. For example, the age span is of 5 years while the period span is of 3 years. In such a case, a span of the least multiple is needed, such as a span of 15 years, a multiple of both age and period spans. By taking the least multiple, the data may be further "collapsed" into a table with the least multiple span for both the age and period, which will make the age, period, and cohort pattern under the new span of the least multiple. Then the assumptions of the APC models become valid and the estimation procedure may carry out properly.

4.2 Complexity in Full Age-Period-Cohort Models

As shown in the last section, the age, period, and cohort factors have interesting relationship among them, and present complex patterns. Such patterns further induce complexity in the statistical models with the age, period, and cohort as covariates. In this section, I will present the model complexity and illustrate the challenging parameter identification problem in the age-period-cohort models. I first review linear regression models and discuss how linearly dependent covariates may induce the parameter identification problem. I will then demonstrate the linear dependence among the age, period, and cohort, and further illustrate the identification problem with the US lung cancer mortality data.

4.2.1 Regression with Linearly Dependent Covariates

Linear regression models (or linear models) are often written in the following form

$$Y_i = \mu + b_1 x_{i1} + b_2 x_{i2} + \ldots + b_k x_{ik} + \varepsilon_i, \quad i = 1, \ldots, n,$$

where Y_i is the response variable of subject i. $x_{i1}, \ldots, x_{ik}$ are the values of k predictors (or covariates) of subject i with the corresponding coefficients $b_1, \ldots, b_k$, respectively, as model parameters. μ is the model intercept, and ε_i is the random error of subject i, which is often assumed to be independent and follow a Gaussian distribution $N(0, \sigma^2)$ with mean 0 and common variance σ^2.

Often, the sample size n is assumed to be larger than the number of covariates k, $n > k$. To find the parameter estimates, the Gauss's least-squares (L-S) method minimizes the following sum of squares of the errors

$$\min_{(\mu,b_1,\ldots,b_k)} \sum_{i=1}^{n} [y_i - (\mu + b_1 x_{i1} + b_2 x_{i2} + \ldots + b_k x_{ik})]^2, \qquad (4.1)$$

where y_i is the observed value of the response of subject i.

To explain the issue of the linearly dependent covariates, I write the above linear model in the matrix form

$$\boldsymbol{Y} = X\boldsymbol{b} + \varepsilon, \qquad (4.2)$$

where $\boldsymbol{Y}$ is an $n \times 1$ vector of response, $\boldsymbol{b} = (\mu, b_1, \ldots, b_k)^T$ is a $(k+1) \times 1$ vector of model parameters, X is the regression matrix of dimension $n \times (k+1)$ corresponding to the model parameters, and ε is an $n \times 1$ vector of independent random errors with mean $\boldsymbol{0}$ and variance-covariance matrix $\sigma^2 I$, where I is an identity matrix of dimension $n \times n$ with entry 1 on the main diagonal and entry 0 off the main diagonal.

The least-squares (4.1) may be written as

$$\min_{\boldsymbol{b}} [(\boldsymbol{Y} - X\boldsymbol{b})^T (\boldsymbol{Y} - X\boldsymbol{b})]$$

and the L-S estimate of the parameters $\boldsymbol{b}$ is given by

$$\widehat{\boldsymbol{b}} = (X^T X)^{-1} X^T \boldsymbol{Y}. \qquad (4.3)$$

By the linear model theory, the L-S estimator is valid when the covariates of the model (4.2) are not linearly dependent. Linearly dependent covariates make the matrix X less than full rank because its column vectors are linearly dependent, and the matrix $X^T X$ is singular with multiple generalized inverse $(X^T X)^-$. In addition, there exists some nonzero vector $\boldsymbol{v}$ such that $X\boldsymbol{v} = \boldsymbol{0}$, a zero vector. The vector $\boldsymbol{v}$ is in the null space of the matrix X, such as the eigenvector of the matrix $X^T X$ with eigen value 0. Furthermore, for an arbitrary real number $t \in \mathbf{R}$, $X t \boldsymbol{v} = t X \boldsymbol{v} = \mathbf{0}$. Hence if $\boldsymbol{u}$ is an estimator of the linear model (4.2), it satisfies the normal equation $(X^T X) \boldsymbol{u} = X^T \boldsymbol{Y}$. Then $\boldsymbol{u} + t\boldsymbol{v}$ is also an estimator of the model since it also satisfies the normal equation $(X^T X)(\boldsymbol{u} + t\boldsymbol{v}) = (X^T X)\boldsymbol{u} + \mathbf{0} = (X^T X)\boldsymbol{u} = X^T Y$, indicating the existence of multiple estimators, one for each value of t, while t may take any real number.

4.2.2 Age-Period-Cohort Models and Complexity

When neither age-period nor age-cohort model provides a good fit to the APC data in the rectangular table, the following full age-period-cohort model may be applied to provide a better fit. The model is often written as the multiple classification model, as it was named in Kupper et al (1985).

$$Y_{ij} = \mu + \alpha_i + \beta_j + \gamma_k + \varepsilon_{ij}, \quad i = 1, \ldots, a, \;\; j = 1, \ldots, p, \;\; k = a - i + j, \;\; (4.4)$$

where Y_{ij} is the response variable in cell (i, j) in the i-th row and j-th column. α_i, β_j, and γ_k are the i-th age, j-th period and k-th cohort effects, respectively. Although it seems that the full APC model (4.4) takes the form as the ANOVA model with discrete age, period, and cohort factors, the model fitting is complex since the linear dependence among the levels of the age, period, and cohort factors induces a singular design matrix of model (4.4). Thus there exist multiple estimators of the model parameters as discussed in the last subsection. Although all multiple estimators yield the same fitted values, each set of parameter estimates generates specific trends in the age, period and cohort, which varies from set to set, making it difficult to determine which set of parameter estimates is accurate and thus may be used for data analysis and statistical inference.

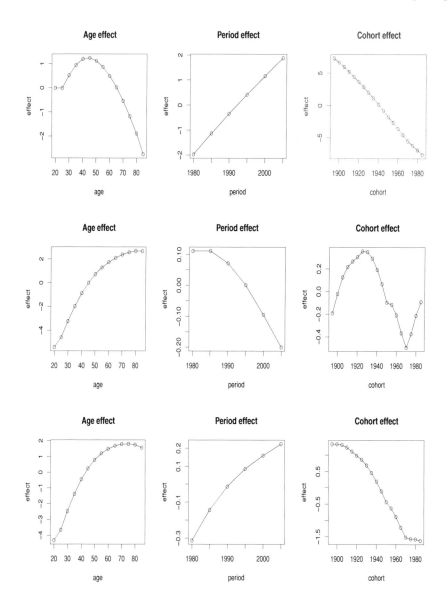

FIGURE 4.1

Plot of parameter estimates of the full APC model of the lung cancer mortality rate among US males. Top panels: assuming identical effects of the two youngest age groups, or $\alpha_{20} = \alpha_{25}$; Middle panels: assuming identical effects of the two earliest periods, or $\beta_{1980} = \beta_{1985}$; Bottom panels: assuming identical effects of the two oldest cohorts, or $\gamma_{1895} = \gamma_{1900}$.

To illustrate the parameter identification problem, I take the following steps.

First, I illustrate the null vector for the singular design matrix of the model (4.4) of an APC data in a 3×3 table. As shown below, the design matrix multiplied by the null eigenvector equals to a zero vector.

$$
\begin{pmatrix}
1 & 1 & 0 & 1 & 0 & 0 & 0 & 1 & 0 \\
1 & 1 & 0 & 0 & 1 & 0 & 0 & 0 & 1 \\
1 & 1 & 0 & -1 & -1 & -1 & -1 & -1 & -1 \\
1 & 0 & 1 & 1 & 0 & 0 & 1 & 0 & 0 \\
1 & 0 & 1 & 0 & 1 & 0 & 0 & 1 & 0 \\
1 & 0 & 1 & -1 & -1 & 0 & 0 & 0 & 1 \\
1 & -1 & -1 & 1 & 0 & 1 & 0 & 0 & 0 \\
1 & -1 & -1 & 0 & 1 & 0 & 1 & 0 & 0 \\
1 & -1 & -1 & -1 & -1 & 0 & 0 & 1 & 0
\end{pmatrix}
\times
\begin{pmatrix}
0 \\ -1 \\ 0 \\ 1 \\ 0 \\ -2 \\ -1 \\ 0 \\ 1
\end{pmatrix}
=
\begin{pmatrix}
0 \\ 0 \\ 0 \\ 0 \\ 0 \\ 0 \\ 0 \\ 0 \\ 0
\end{pmatrix}
$$

Second, I illustrate the multiple sets of parameter estimates with the lung cancer mortality rates among US males. Figure 4.1 displays three sets of curves by the parameter estimates for the effects of the age, period, and cohort. It is clearly shown that these three sets display different trends, though they have exactly the same fitted values in fitting the model (4.4).

It is well known that lung cancer mortality increases with age and thus the decreasing age trend from age 40 to 85 in the top-left panel is not supported. While the increasing age trend in the middle-left and bottom-left panels makes sense, the period trend differs largely between the middle and bottom panels. Overall, the multiple sets of the parameter estimates and the trend estimation often lead to confusion and misinterpretation in the literature. This is the identification problem (or the identifiability problem) in the APC analysis that has been observed during the last four decades, and has been regarded by many quantitative scientists, including biostatisticians, demographers, economists, epidemiologists and sociologists, as an open problem yet impossible to resolve.

I will elaborate with more details in the next chapter why this problem has been challenging and deemed unsolvable. I will also review a number of approaches in the literature that have been studied to deal with the identification problem. I will then provide a novel approach to resolving the identification problem in Chapter 6, demonstrate how to use the novel method to analyze data in Chapter 7, and further provide detailed justification based on statistical theory in Chapter 8.

4.3   R Programming for Generating the Design Matrix for APC Models

Unlike the single factor or two factor models, which are special cases of fixed effect ANOVA models, the APC model (4.4) has a special form and thus requires a special design matrix. An **R** function apcmat generates the design matrix for a given APC data of dimension $a \times p$. For example, the design matrix in section 4.2.2 was generated with the **R** command apcmat(3,3). The eigen values and eigenvectors of the design matrix of a given APC data of dimension $a \times p$ can be generated with the commands below.

```
x=apcmat(a, p)
eigen(t(x)%*%x)$value
eigen(t(x)%*%x)$vector
```

4.4   Suggested Readings

The identification problem has been observed and reported since the 1970s and has received much attention in the literature. Many papers made observations and numerical demonstrations of the identification problem, while some others also made further explorations, trying to understand the problem deeply using mathematical statistics techniques, such as constraint methods, eigen analysis, estimable functions, etc. See Fienberg and Mason (1985), Holford (1983, 1991), Kupper et al (1983, 1985), Mason and Smith (1985), Robertson and Boyle (1986), Clayton and Schifflers (1987b), Rodgers (1982) and Fienberg (2013) for details. Interested readers may find a large amount of publications in statistics, public health, and social sciences.

4.5   Exercises

1. Why is the equal span of the age and period an important issue?

2. What would happen if the unequal spans in the age and period are ignored and an age-period-cohort model is fitted to the data assuming equal span?

3. Can one simply collapse multiple cells into one to ensure the equal span in the age and period? What information would be lost in such a practice? Explain with the retirement accounts data in Table 1.4.

4.* Can you come up with a procedure to code the special cohort effect in an unequal span setting?

---

* Difficult exercises with an asterisk are meant for graduate students in biostatistics.

5

Age-Period-Cohort Models — The Identification Problem and Approaches

The linear dependence among the age, period, and cohort has been shown to induce the identification problem in model parameter estimation. Since it is crucial to identify the trends in the age, period, and cohort of the events under investigation, such as chronic diseases (e.g. various types of cancer, or cardiovascular diseases) in public health studies or crimes (e.g. homicide, suicide) in social behavior studies, many methods have been studied in the literature to address the identification problem. Among them, two popular ones play a major role as I show in this chapter, while the others provide alternative approaches to the problem and may achieve reasonable estimation. However, the challenges presented in the identification problem make it exceedingly difficult to resolve with the classical statistical reasoning and approaches.

5.1 The Identification Problem and Confusion

The identification problem causes tremendous confusion in the literature. Many believe that the existence of the multiple estimators implies that there exist no true parameter values of the APC models (4.4) for a given data set, simply because the same data set may be generated by many sets of parameters that present different trends in the age, period, and cohort, as shown in Figure 4.1. Hence more information is needed to determine the parameter values, yet such information must come from the external knowledge of the diseases or events under investigation. This belief motivated various approaches to exploring further information in the APC studies. One typical approach is to collect more data or acquire more knowledge about the study. It may be achieved by setting an extra constraint on the parameters based on the investigator's prior knowledge of the event under investigation. However, it is well known that the parameter estimates depend on the specified constraint, different constraint leads to different parameter estimates and may present different trend. The approach may also be achieved by collecting individual subject's birth record so that the model may have more detailed cohort information, which may potentially help to determine the parameters. But this also bears extra burden of collecting a large number of individual records in a national study. Another approach is to consider the properties of the estimates and ensure the parameter estimates to perform well. It may be achieved by identifying a special type of functions of the parameters, called the estimable functions. The estimable functions do not vary with the constraint specified, and thus have unbiased estimates. It may also be achieved by sensitivity analysis so that special parameter estimates are obtained to ensure robust parameter estimation.

Among the various approaches, two of them have received much attention. One is to specify an extra constraint on the parameters of the model so that a unique set of parameter estimates can be identified, and a secular trend can be uniquely determined. The other is to identify estimable functions so that the parameter estimates may be determined and not subject to estimation bias. I elaborate both of them in detail in the next two sections, and show that they play a crucial role in addressing the identification problem and provide key clues to the resolution to the identification problem. I also review other approaches that have been studied in the literature.

Although it has been a general belief in the APC community for a long time that the APC models, such as equation (4.4), do not have true parameter values for any given data, it is worthwhile to note that such parameter identification problem is model-specific but not data-specific. For a given data set, whether the true age estimates can be estimated accurately is unknown without searching for all possible approaches. For example, the age-period (AP) model yields a unique set of parameter estimates, and may lead to unbiased estimates of the age effects if the true cohort effects are 0 and not present. In such a case, the data has true parameter values, which can be accurately estimated with the AP models, but difficult to estimate with the APC models (4.4). Furthermore, a statistical model, in particular, the APC model (4.4), only yields parameter estimates for a data set of finite sample, but never yields true parameter values. The true values may only be revealed by consistent parameter estimators as the sample size increases to infinity, following the law of large numbers in statistical theory. Hence the difficulty in and impossibility of determining a unique set of parameter values by a model, such as model (4.4), for a given finite data set may not imply that there exist no true parameter values, because a data set only presents a random sample and consequently the parameter estimates of a model only constitute at the best an approximation to the parameters, but not the true parameters themselves. This implicates that a study of the large sample behavior of the estimators is needed to reveal the existence of the true parameters of the APC models (4.4). I will provide the rationale of such studies in Chapter 6 so that it becomes easy to understand to the readers who do not have background in statistical theory and are not interested in it.

5.2   Two Popular Approaches to the Identification Problem

It is known that the APC identification problem is induced by the linear dependence among the covariates of model (4.4), or the lack of full rank of the design matrix of the model, which yields multiple estimators. A straightforward approach is to achieve a unique estimator, which makes it easy for parameter estimation and interpretation. This can be achieved by various methods, as discussed below.

5.2.1   Constraint Approach

Specifying a constraint is the first choice among all possible methods that may help to resolve the problem. Since there are many ways to specify a constraint, and each constraint may lead to a set of estimates different from others, it makes the selection of the constraint crucially important. The selection of a constraint based on prior knowledge of the events under investigation often becomes the first priority. The rationale of specifying an extra constraint is essential to resolving the identification problem.

Rationale of the Constraint Approach *Accurate parameter estimates may be yielded by specifying a constraint that is satisfied by the true parameter values.*

This rationale is fundamental in dealing with the existence of multiple estimators and the identification problem, but quickly leads to a paradox, as observed in the literature. In order to specify a constraint that is satisfied by the true parameter values, one needs to know the true values first, which is impossible before solving the problem with accurate estimation. This paradox seems impossible to resolve, and consequently, alternative approaches are pursued relying on prior knowledge of the events under investigation. For example, most cancer mortality rate increases with age, thus an upward trend is likely to be more accurate than a downward trend and is acceptable in studies of cancer mortality rate, etc. As shown in the US male lung cancer mortality study in Figure (4.1), the increasing-then-decreasing age trend in the top-left panel is not acceptable, while the increasing age trends in the middle and bottom panels make more sense. However, this does not completely determine the parameter estimates, as shown in the other two panels with different period trends. Thus one often relies on the investigator's subjective assumption to specify an extra constraint. In the literature, it is fairly straightforward to specify an equality constraint, such as $\alpha_1 = \alpha_2$, based on the assumption that there is no major difference in lung cancer mortality in early ages and consequently the two youngest age groups have identical effects. However, such a fundamental principle cannot yield reasonable results in practice, as it has been pointed out that constraints based on seemingly reasonable assumptions often lead to unacceptable trend estimation (Kupper et al 1985). As shown in Figure 4.1, a reasonable assumption on the identical age effects in the early ages on male lung cancer mortality leads to an unacceptable increasing-then-decreasing age trend in the top-left panel. I will explain why such seemingly reasonable constraint often leads to unacceptable trend estimation in Chapter 6 and provide full detailed theoretical explanations in Chapter 8.

Although the above discussion and prior studies in the literature seem to demonstrate that the paradox is impossible to resolve, I point out that reasonable parameter estimates may be achieved with alternative models or methods, and thus the specification of the constraint that is almost certainly satisfied by the true parameter values is not only possible but also not difficult to achieve. I will give more details on how to specify such constraints in the later chapters.

I also would like to make a key point about the rationale of the constraint approach. Although it seems exceedingly difficult, if not impossible, to specify a constraint that is fully satisfied by the model parameters, the rationale provides theoretical justification on assessing the true parameters: a constraint satisfied by the set of parameters would yield unbiased estimates, while a constraint not satisfied by the parameters would yield biased estimation. Following the rationale, I will provide theoretical justification on a novel approach to the identification problem in the next few chapters.

5.2.2   Estimable Function Approach

Identifying estimable functions is one of the key approaches to resolving identification problems whenever multiple sets of parameter estimates exist and a decision on the parameter estimation is difficult to make. For example, in the one-way ANOVA model, different side conditions may be specified in order to determine a set of parameter estimates, leading to multiple sets of estimates. Bias is thus incurred in the parameter estimation by setting a side condition since the estimates depend on and may vary largely with the side condition, which is often chosen based on the investigator's individual preference. Estimable functions are those functions of the model parameters that do not vary with such side conditions, and thus have unique unbiased estimates because the estimates do not vary with the side condition either. The well known estimable functions in the one-way ANOVA models are the contrasts $l^T b$ of the parameters b, where the coefficients $l = (l_1, \ldots, l_k)^T$ satisfy $\sum_1^k l_j = 0$ in a k group one-way ANOVA model. It can be shown that the value of a contrast does not vary with the side condition, or any side condition yields the same value of the contrast.

Given the property of the unbiasedness, identifying the estimable functions may potentially provide a resolution to the identification problem in the APC models. In particular, an estimable function that completely determines the parameter values, or the trend in the age, period, and cohort, would be a resolution. The rationale of the estimable function approach is as follows.

Rationale of the Estimable Function Approach *Estimable functions have unbiased estimates, which do not depend on the constraint that is used to determine the estimates.*

Given that the estimable functions possess such desirable properties, much effort has been made in the literature in searching for the estimable functions (Rodgers 1982, Smith, Mason and Fienberg 1982, Kupper et al 1985, Holford 1985, 1991).

It has been observed in the literature that the nonlinear characteristics of the trend are estimable, such as curvatures of age, period, and cohort trends, as shown in Figure 5.1, but the linear trend, or the overall slope, varies with the constraint or the estimates. This observation leads to the conclusion that the identification problem is unsolvable (Kupper et al 1985, Mason and Fienberg 1985, Clayton and Schifflers 1987b, Holford 1991, etc).

It is worthwhile to note the following. First, the curvature and other non-linear characteristics do not vary with the curve estimated through the constraint. However, since the linear trend may vary largely, the nonlinear characteristics may be dominated by the linear trend and thus may be difficult to observe, as shown in Figure 4.1, where the nonlinear curvature in the cohort trend is readily observable in the middle-right and the bottom-right panels, but difficult to observe in the top-right panel. The largely different scales of

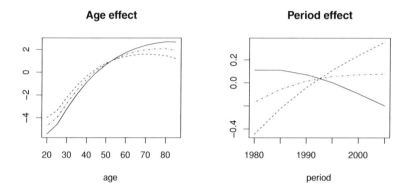

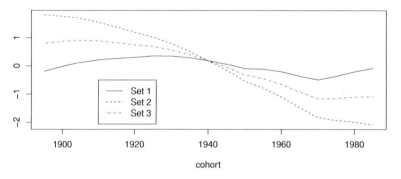

FIGURE 5.1

Plot of 3 sets of parameter estimates of the full APC model on the lung cancer mortality rate among US males, indicating the invariant curvature and other nonlinear characteristics.

the cohort plot in these three panels may help to explain why this is the case. Second, strictly speaking, estimable functions are, by definition, linear functions of the model parameters according to linear model theory, such as the contrasts. However, in the APC analysis, the concept of estimable functions is further extended to nonlinear functions that do not vary with the constraint on the model parameters (Rodgers 1982, Smith, Mason and Fienberg 1982, Kupper et al 1985, Holford 1983, 1991) because the focus is to search for parameter estimates that do not vary with the constraint, as discussed in the previous section.

It is also of great interest to notice that a sufficient condition for the identification of such an estimable function was provided in Kupper et al (1985), but was criticized in a commentary paper by Holford (1985) for being oversimplified given the complexity of the identification problem. The sufficient condition states that any linear function of the model parameters $l^T b$ is estimable if l is perpendicular to the null vector of the singular design matrix X of the APC model (4.4). It provides a simple and sufficient condition that is easy to examine through straightforward calculation, but did not receive attention in the literature.

5.3 Other Approaches to the Identification Problem

In addition to the above two popular approaches in the literature, other approaches have also been studied, which may or may not directly address the identification problem, including a shrinkage method (Osmond and Gardner 1982), a drift model (Clayton and Schifflers 1987), a graphical method (Robertson and Boyle 1998), a ridge estimator method and its limit (Fu

2000), the age-period-cohort-characteristic model (O'Brien 2000), the smooth-ing method (Heuer 1997, Fu 2008), a partial least-squares method (Tu et al 2011, 2012), and a series of Bayesian models (Berzuini and Clayton 1994, Yang 2006, Schmid and Held 2004, Schmid et al 2007), to name a few. These methods provide alternative approaches to the identification problem and may yield a unique estimator of the parameters. However, they may or may not address the key issue of the identification problem. Very often, an alternative method may yield a unique estimator and consequently a unique set of param-eters and secular trend may be completely determined. However, the crucial component of providing justification as to why such an alternative approach addresses the identification problem is often left unattended. For example, Osmond and Gardner (1982) studied a shrinkage approach by noticing that taking the estimator among the multiple ones of model (4.4) that has the smallest distance from the estimators of all three models, the AP, AC and PC models, may help to determine a unique set of parameter estimates. Geo-metrically, the estimator of each of the three marginal models of AP, AC and PC lies on a hyperplane of the multi-dimensional parameter space composed by the coordinates of the APC estimates, leading to a unique estimation of the parameters. But such an approach does not explain why, geometrically, the estimator having the smallest distance to all three estimators (thus the name shrinkage estimator) can be justified and what properties such an esti-mator possesses. Another example is the ridge estimator studied in Fu (2000). It is well known that ridge penalty yields a unique and robust estimator in regression models with a potential issue of collinearity, i.e. the predictors are close to linearly dependent. However, why such a ridge estimator may help to address the identification problem still remains unknown.

It is known that any attempt to avoid these major issues will not address the problem properly and may only provide alternative approaches, leaving the identification problem unsolved. Interested readers are referred to a review paper (Fu et al 2017).

5.4 Suggested Readings

An overview of the identification problem can be found in a number of review papers, including Kupper et al (1985), Clayton and Schifflers (1987a, 1987b), Robertson and Boyle (1998a,1998b), Holford (1991). A number of alternative approaches can be found in the literature, including a shrinkage estimator (Osmond and Gardner 1982), an individual record method (Robertson and Boyle 1986), a smoothing age-period-cohort model (Heuer 1997), a ridge estimator method and its limit (Fu 2000), an age-period-cohort-characteristic model (OBrien 2000), a Bayesian method (Schmid and Held 2004, 2007), a smoothing cohort model (Fu 2008), a partial least-squares method (Tu et al 2011, 2012), and a sensitivity analysis method (Fu 2016).

5.5 Exercises

1.* What are the estimable functions in the APC models (4.4), excluding the nonlinear functions of the parameters?

2.* How to find all estimable functions for a given model, such as model (4.4)?

3.* Given that the slopes of the trends in the age, period, and cohort vary with

---

* Difficult exercises with an asterisk are meant for graduate students in biostatistics.

the constraint, does it imply that there does not exist an estimable function of the parameters of model (4.4) that determines the slope of the trend?

6

Intrinsic Estimator, the Rationale and Properties

As I discussed in Section 5.3, no matter how well an alternative method performs in estimating age, period, and cohort effects, and estimating the secular trend, the identification problem still causes confusion and needs to be addressed. Specifically, the issues of whether there exists a set of true parameters of the model, and if so, how to estimate them, need to be addressed. In this chapter, I address the identification problem and provide a straight answer. I show that contrary to the belief in the literature, the identification problem can be resolved, leading to the clarification of the long-term confusion. Furthermore, the recently developed intrinsic estimator and its theoretical properties provide guidance to the applied statisticians and quantitative scientists who are interested in using the method to analyze APC data in practice. I first provide a complete structure of the multiple estimators of the APC model (4.4). This structure of estimators leads to the discovery of the estimable function that determines completely the trend in the age, period, and cohort. I further study the properties of the intrinsic estimator defined based on the estimable function. I then provide an outline of the justification, and a heuristic explanation of why this method works while others may not. I will demonstrate the method with the lung cancer mortality data and the retirement accounts data in the next chapter with **R** programming, and provide rigorous theoretical justification with full details based on the statistical large sample theory in Chapter 8.

6.1 Structure of Multiple Estimators of Age-Period-Cohort Models

I have shown in the previous chapter that the multiple estimators of the age-period-cohort model (4.4) lead to the difficult identification problem. To address the problem, one needs to study the multiple estimators and achieve a thorough understanding. A structure of the multiple estimators is given in Fu (2000)

$$\widehat{b} = B + tB_0 \,, \tag{6.1}$$

where $b = (\mu, \alpha_1, \ldots, \alpha_{a-1}, \beta_1, \ldots, \beta_{p-1}, \gamma_1, \ldots, \gamma_{a+p-2})^T$ is the vector of model parameters, and B_0 is the unit null eigenvector of the singular design matrix X of model (4.4). B is the vector in the parameter space perpendicular to B_0 and satisfies the normal equations. t is an arbitrary real number. Since the matrix X is only 1-less than its full rank, it has one eigenvalue 0 and thus a unique unit null eigenvector B_0. As shown in Figure 6.1, since t is arbitrary, the multiple estimators correspond to the multiple values of t. Consequently, there are infinite and uncountable multiple estimators.

The null eigenvector has a closed form (Kupper et al 1985)

$$v = (0 \, A \, P \, C)^T, \tag{6.2}$$

where row vectors A, P and C are given by

$$A = \left(1 - \frac{a+1}{2}, 2 - \frac{a+1}{2}, \ldots, a - 1 - \frac{a+1}{2} \right),$$

$$P = \left(\frac{p+1}{2} - 1, \frac{p+1}{2} - 2, \ldots, \frac{p+1}{2} - (p-1) \right),$$

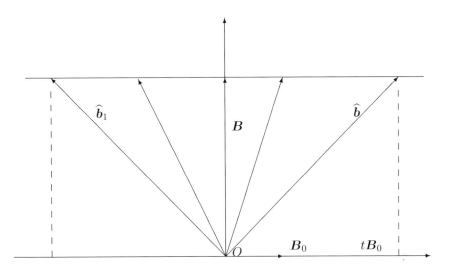

FIGURE 6.1
Structure of the multiple estimators of the age-period-cohort model (4.4).
A vector from the origin to any point on the top line is an estimator of the
parameters. It is decomposed to the vertical vector $\boldsymbol{B}$ and a horizontal vector
$t\boldsymbol{B}_0$, a multiple of the null eigenvector $\boldsymbol{B}_0$ along the horizontal axis.

$$C = \left(1 - \frac{a+p}{2}, \, 2 - \frac{a+p}{2}, \ldots, a+p-2-\frac{a+p}{2}\right).$$

Then the unit null vector $\boldsymbol{B}_0 = \boldsymbol{v}/\|\boldsymbol{v}\|$, where $\|\boldsymbol{v}\|$ is the L_2 norm of the

vector $\boldsymbol{v}$,

$$\|\boldsymbol{v}\| = \sqrt{\sum_{i=1}^{a-1}\left(i - \frac{a+1}{2}\right)^2 + \sum_{j=1}^{p-1}\left(\frac{p+1}{2} - j\right)^2 + \sum_{k=1}^{a+p-2}\left(k - \frac{a+p}{2}\right)^2}.$$

6.2 Intrinsic Estimator: Unbiased Estimates and Other Properties

The structure of the estimators (6.1) makes it easy to achieve a thorough

understanding of the multiple estimators and their behavior. Furthermore, the

structure in Figure 6.1 provides further insight to identifying the estimable functions. It shows that among all estimators, B is uniquely in the center and is perpendicular to the null vector B_0. Its expectation surely satisfies the condition of the estimable function by Kupper et al (1985). It can be further proved rigorously that the expectation $E(B)$ is estimable (Fu 2016); see Chapter 8 for details of the proof. Since it determines a unique set of parameters, it is the estimable function that has been searched for but was unfortunately claimed to not exist in the literature.

The discovery of the estimable function raises an important question: Whether one may use the estimable function and its corresponding estimator to analyze APC data. It turns out that not only can it be used to analyze APC data, but it also provides consistent estimation of the model parameters. That is to say, this method provides parameter estimates converging to the true values as the sample size increases to infinity, hence yielding the best possible estimation among the infinitely many estimators. The unbiased estimator of the estimable function is called the intrinsic estimator, which was first discovered in Fu (2000) as the limit of the ridge estimator when the penalty tuning parameter diminishes to 0.

Properties of the Intrinsic Estimator

The intrinsic estimator possesses the following properties, as studied in Fu (2016).

1) Its expectation is an estimable function that determines the parameters and the trend, hence it represents invariant properties of the multiple estimators and is not subject to arbitrary constraints.

2) It yields robust trend estimation with slight changes in the data, such as having either one less age group or one less period.

3) It is the limit of the ridge estimator as the penalty diminishes to 0, thus inheriting properties of the ridge estimator, such as small variance, etc.

4) From a Bayesian point of view, it is the posterior estimator with a non-informative prior.

5) It converges to the true parameter values in probability as the sample size diverges to infinity — the desirable consistency of parameter estimation.

6) It achieves the square-root consistency and asymptotic normality, thus converges to the true parameters at the fastest speed, and follows approximately a normal distribution.

Given the above properties, the intrinsic estimator no doubt performs well in data analysis. In particular, since the intrinsic estimator is consistent, i.e., the estimates of the age effects converge in probability at the fastest speed to the true age effects as the number of periods, and thus the sample size, tends to infinity, while the other estimators do not, it performs the best among the infinitely many estimators and yields accurate estimates of age, period, cohort effects and their secular trends.

I now provide details of a robust parameter estimation study through a sensitivity analysis and summarize the asymptotic properties of the intrinsic estimator and the multiple estimators based on the large sample theory.

6.3   Robust Estimation via Sensitivity Analysis

It is known from previous studies that the multiple estimators behave differently. Some may be sensitive to slight changes in the data while some others are robust. From the application point of view, robust trend estimation is always preferred, as slight changes of data are always expected in practice, such as having one more age group or period. If such slight changes of data

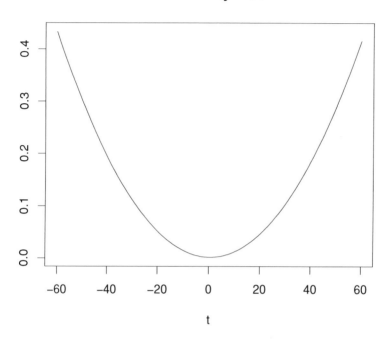

FIGURE 6.2
The sensitivity function $S(t)$ of lung cancer mortality rate among US females. The first age group (20–24 years old) is excluded because of missing data in the log-transformed rate.

induce a dramatic change in the trend estimation, the estimated trend will be of little use in practice.

To study the sensitivity of the multiple estimators (6.1), one may quantify the sensitivity of an estimator, corresponding to a value of t, using the change of the trend estimates. Since the APC data lies in a rectangular table, a slight change of the data may mean adding or dropping one row (age group) or one column (period). In doing so, the number of age groups may change while the number of the periods remains the same, or vice versa. And the number of diagonals (or birth cohorts) changes with either the number of rows or that of columns. Hence, the quantification of the change of the trend estimates is defined separately for the changes in rows and the changes in columns.

Specifically, the change of the trend estimates may be defined to be the Euclidean norm of the difference in the estimates. I introduce some useful notations. For each fixed t, let $\widehat{\boldsymbol{b}}(t, age)$ and $\widehat{\boldsymbol{b}}(t, per)$ be the age and period effect estimates based on the full data in an $a \times p$ table, respectively. Let $\widehat{\boldsymbol{b}}(t, age, -1)$ and $\widehat{\boldsymbol{b}}(t, age, -p)$ be the age effect estimates with the first period and the last period dropped from the full data, respectively. Similarly, let $\widehat{\boldsymbol{b}}(t, per, -1)$ and $\widehat{\boldsymbol{b}}(t, per, -a)$ be the period effect estimates with the first age group and the last age group dropped from the full data, respectively. Define a sensitivity function $S(t)$ of the estimator $\widehat{\boldsymbol{b}}(t) = \boldsymbol{B} + t\boldsymbol{B}_0$ with a given t as follows.

$$
\begin{aligned}
S(t) &= \frac{1}{2a}\left[||\widehat{\boldsymbol{b}}(t, age, -1) - \widehat{\boldsymbol{b}}(t, age)||^2 + ||\widehat{\boldsymbol{b}}(t, age, -p) - \widehat{\boldsymbol{b}}(t, age)||^2\right] \\
&+ \frac{1}{2p}\left[||\widehat{\boldsymbol{b}}(t, per, -1) - \widehat{\boldsymbol{b}}(t, per)||^2 + ||\widehat{\boldsymbol{b}}(t, per, -a) - \widehat{\boldsymbol{b}}(t, per)||^2\right],
\end{aligned}
$$

$$(6.3)$$

where $|| \cdot ||$ is the Euclidean norm of a vector.

The cohort effects are not included in the calculation of the sensitivity function $S(t)$ since dropping either one row or one column changes the number of diagonals (cohorts), making it difficult to assess the change of the cohort effects.

Fu (2016) proves that the sensitivity function $S(t)$ behaves like a quadratic function for large values of $|t|$. It can be observed in Figure 6.2 that the sensitivity function $S(t)$ is approximately quadratic and diverges to infinity as $|t| \to \infty$. It also achieves a minimum at a finite value close to 0, which further indicates that the intrinsic estimator (corresponding to $t = 0$) induces little sensitivity and yields robust estimation.

Intensive simulation studies in Fu (2016) show that robust estimation of the parameters is always achieved near $t = 0$, which implies that the intrinsic estimator (corresponding to $t = 0$) presents the least sensitivity and yields robust trend estimation, making it desirable in data analysis.

6.4   Summary of Asymptotic Properties of the Multiple Estimators

I now summarize the asymptotic results of the multiple estimators, aiming to explain why the intrinsic estimator can address the identification problem and provide the best parameter estimation, and why the constraint method does not often yield reasonable analysis results.

For the reader's convenience, I summarize the major results in this section, and provide all detailed theoretical justification with rigor in Chapter 8. Readers who are not interested in the large sample theory may read this section to have an understanding of the overall rationale without paying attention to the rigorous technical details of the theory. Readers who wish to know more details may read Chapter 8, which contains large amounts of statistical theory and is suitable for statisticians and graduate students who have a solid statistical background and are interested in learning the justification of the intrinsic estimator method.

First, I explain why the asymptotics study requires large p. To study the asymptotic properties of the estimators of a statistical model, such as the logistic regression, one needs to let the sample size go to infinity to study the consistency and efficiency of the parameter estimates. Notice that even if a model has no linearly dependent covariates, the unique parameter estimates based on a sample of data only provide estimation of the parameters, but not the true parameter values themselves. Only as the sample size diverges to infinity may the true parameter values be revealed by consistent estimates, hence the asymptotic studies are always needed for accurate estimation and sound inference. However, notice that the APC data only has virtually one data point in each cell (an archived number) and the total sample size is ap for an $a \times p$ table. To let the sample size go to infinity, one needs to let either the number of rows (age groups) or the number of columns (periods) go to infinity, or both. Since the human lifespan is about 100 years, it is not practical to have $a \to \infty$. Having $p \to \infty$ means to have more periods in the study design, which is meaningful and achievable in theory. Hence the asymptotic study is conducted as $p \to \infty$.

I would further like to make the following two key points in the study of the asymptotics. 1) Because the numbers of periods and cohorts diverge to infinity as $p \to \infty$, one may only study the asymptotic behavior of the model intercept and age effect estimates. Thus, a profile likelihood approach is employed, which takes the intercept and age effect parameters as the primary parameters, and the period and cohort effects as parameters of secondary interest or as nuisance parameters. The asymptotics are studied on the behavior of the primary parameters only. 2) Once the estimates of the primary parameters are shown to be consistent, the parameters of the secondary interest or the profiles may be shown to yield consistent estimation assuming the profile function (or the model itself) remains to be true.

To study the asymptotics, one needs to make assumptions on the smoothness of the likelihood functions in the following regularity conditions. The mild regularity conditions here require 1) the likelihood function of the distribution of the random response variable is in the exponential distribution family and is smooth enough to have continuous second order derivatives and uniformly bounded first order derivatives; 2) the period and cohort effects are uniformly bounded with respect to p. The first condition is on the likelihood function, and is very mild as the most popular distributions in the exponential family, such as the binomial distribution or the Poisson distribution, satisfy the condition. The second condition is on the boundedness of the period and cohort effects as the primary interest is not in studying diverging effects of the period or the cohort, which may blow up as $p \to \infty$.

For convenience, I introduce the following notations for the asymptotic studies. Let $\boldsymbol{\theta} = (\mu, \alpha_1, \ldots, \alpha_{a-1})^T$ be the primary parameters and $\boldsymbol{\xi} = (\beta_1, \ldots, \beta_{p-1}, \gamma_1, \ldots, \gamma_{a+p-2})^T$ be the nuisance parameters. Further let $\widehat{\boldsymbol{\theta}}$ and $\widehat{\boldsymbol{\xi}}$ be the estimates by the intrinsic estimator. The major asymptotic results are summarized below.

Proposition 6.1 *The intercept and age effect estimates by the intrinsic estimator converge in probability to a unique limit $\boldsymbol{\theta}^\infty$ as $p \to \infty$,*

$$\widehat{\boldsymbol{\theta}} \to_p \boldsymbol{\theta}^\infty \quad as \ p \to \infty.$$

Proposition 6.2 *A constraint on the age effects yields biased estimates unless the constraint is satisfied by the limit $\boldsymbol{\theta}^\infty$ of the age effect estimates.*

Given the fact that an age effect constraint yields consistent estimates if it is satisfied by the limit $\boldsymbol{\theta}^\infty$, and yields asymptotically biased estimates if it is not satisfied by the limit, it is clear that the unique limit $\boldsymbol{\theta}^\infty$ functions as the true age effect parameters by the rationale of the constraint approach as discussed in the previous chapter. Hence the unique limit $\boldsymbol{\theta}^\infty$ is regarded as the parameters of true age effects.

Proposition 6.3

1. A constraint on the period effects or cohort effects yields consistent age effect estimates if it is satisfied by the period or cohort effects as a profile function of the true age effects $\boldsymbol{\theta}^{\infty}$.

2. If a constraint is a contrast (i.e. the sum of all coefficients is equal to 0) and is not satisfied by the profile period or cohort effects, it yields asymptotically biased estimates of the age effects.

3. If a constraint is not a contrast and is not satisfied by the profile period or cohort effects, it yields consistent estimates.

The above propositions provide an outline of the theoretical justification for the intrinsic estimator. The following remarks help to make it easy to understand.

Remarks

1) The true parameter values (which can only be revealed by the asymptotic behavior of consistent estimators, and not by finite sample estimates) exist for the APC model (4.4), contrary to the belief in the literature. The conclusion of the nonexistence of the true parameter values was made based on the observation of the multiple estimators and the indetermination of the parameter estimates, but not based on rigorous theoretical study.

2) The rationale of the constraint approach in the previous chapter is crucial to the existence and discovery of the true parameter values. Proposition 6.2 states that $\boldsymbol{\theta}^{\infty}$ behaves as such true parameter values, i.e. a constraint yields consistent estimation if it is satisfied by $\boldsymbol{\theta}^{\infty}$, and yields biased estimation otherwise. This indicates that the unique limit $\boldsymbol{\theta}^{\infty}$ are the true parameter values, which not only exist but can also be revealed through the limit of the intrinsic estimator $\boldsymbol{B}$.

3) Although multiple estimators exist, they behave differently; some present consistency while the others yield asymptotic bias. One may make a decision of using which estimator to estimate trends in data analysis. The answer is straightforward: use estimators with consistency but not those yielding asymptotic bias so that sound statistical inferences and decisions can be made.

4) Although contrast constraints, such as equality constraint $\beta_1 = \beta_2$ or $\gamma_2 = \gamma_3$, are very popular and easy to implement in practice, it turns out that such constraints almost surely yield biased estimation. This is because one can never expect a constraint to be completely satisfied by the true parameters, hence it is of probability 0 to have a specified constraint satisfied by the true parameters. By Proposition 6.3, such a contrast constraint yields asymptotic bias. Even a tiny bit off from the true parameter values, an equality constraint may induce a non-zero bias, making the estimates deviate from the truth. This helps to explain why a seemingly reasonable constraint on the parameters may often yield unsatisfactory analysis results, as observed by Kupper et al (1985) and many others in the literature.

The above results have been empirically examined and verified through intensive simulation studies in Fu (2016). I do not repeat these simulation results here. Interested readers are referred to Fu (2016) for full details of the simulation studies.

6.5   Computation of Intrinsic Estimator and Standard Errors

6.5.1   Computation of Intrinsic Estimator

Given that the intrinsic estimator is the unique projection of the multiple estimators to the non-null space, which is perpendicular to the unique eigenvector of the design matrix corresponding to the eigenvalue 0, multiple algorithms may be derived to compute the intrinsic estimator. Among them, the simplest one is to remove the null eigenvector and compute the estimator in the non-null space. It is equivalent to compute the parameter estimates following the principal component analysis approach with all principal components of nonzero eigenvalues. The intrinsic estimator can then be obtained by simply converting the estimates on the principal components to the original coordinates by adding 0 to the estimates by the principal components analysis and then rotate the parameter estimates back to the original coordinates of age, period, and cohort effects. I summarize the algorithm below.

Algorithm for the Intrinsic Estimator

1. Form the design matrix X of dimension $ap \times [2(a+p)-3]$ for the parameters

$$(\mu, \alpha_1, \ldots, \alpha_{a-1}, \beta_1, \ldots, \beta_{p-1}, \gamma_1, \ldots, \gamma_{a+p-2})^T.$$

2. Calculate the null eigenvector of matrix $X^T X$ and the orthonormal matrix V composed of all eigenvectors.

3. Regress the response variable Y on the principal components of all non-zero eigen values to obtain the parameter estimates b_0 of dimension $2(a + p - 2)$.

4. Embed the parameter estimates b_0 into a parameter space of dimension $2(a + p) - 3$ by adding a 0.

5. Convert the embedded parameter estimates by the orthonormal matrix V to obtain the intrinsic estimator $B = V\tilde{b}$, where $\tilde{b}^T = (b_0^T, 0)$ is the embedded parameter vector.

6.5.2 Computation of Standard Errors

Standard Errors for Linear Models

For linear model (4.4), the standard errors of the intrinsic estimator can directly be computed following the algorithm for the intrinsic estimator in the previous section.

Algorithm for the Standard Errors

1. Obtain the variance-covariance matrix of the parameter estimates from the regression on the principal components with nonzero eigenvalues, $Var(\boldsymbol{b}_0)$.

2. Compute the variance of the intrinsic estimator $\boldsymbol{B}$ following the linear algebra for the transformation by the matrix V

$$Var(\boldsymbol{B}) = V \begin{pmatrix} Var(\boldsymbol{b}_0) & \boldsymbol{0} \\ \boldsymbol{0}^T & 0 \end{pmatrix} V^T,$$

where $\boldsymbol{0}$ is a vector of dimension $2(a + p - 2)$ with entries 0.

3. The standard errors of the intrinsic estimator $\boldsymbol{B}$ are the square root of the main diagonal elements of the matrix $Var(\boldsymbol{B})$.

Standard Errors for Generalized Linear Models

For the generalized linear models, in particular, the loglinear model (7.1), which I will introduce in the next chapter, the above computation of the standard errors is invalid because the computation relies on the assumption of the approximately normal distribution of the parameter estimates. While the normal distribution is ensured by linear model theory, it is not true in general for the generalized linear models. It relies on the asymptotic normality of the parameter estimates. As I discussed in the previous section, the approximately normal distribution is valid only for the age effect estimates but not for the period and cohort effect estimates. I will introduce the Delta method to compute the variance of the period and cohort effect estimates based on the variance of the age effect estimates, taking the period and cohort effect estimates as profile functions of the age effects. See more details of the Delta method and the variance estimation for the period and cohort effect estimates in Chapter 9.

6.6   Suggested Readings

The structure of the multiple estimators is provided in Fu (2000), in which the intrinsic estimator is also discovered as a limit of the ridge estimator when the ridge penalty diminishes to 0. Fu (2016) provides detailed simulation results in terms of the sensitivity analysis and the asymptotic properties. Technical details of the principal component analysis can be found in Jolliffe (2002).

6.7   Exercises

1.*   Compute the ridge estimator of the linear model (4.4). Let the ridge penalty tuning parameter go to 0 and observe the behavior of the ridge estimates. Do these parameter estimates converge? Can you explain what you observe?

2.* Why does the asymptotic study of the multiple estimators require that the number of columns (periods) p diverge to 0?

3.* Can the large number of individual events in each cell be accounted for the sample size so that the asymptotic studies do not require the number of columns p to diverge to infinity?

4.* Can you find another method to quantify the sensitivity of the estimators? If so, conduct the sensitivity analysis using your novel definition of sensitivity and compare with the sensitivity function (6.3) using the plot shown in Figure 6.2.

---

* Difficult exercises with an asterisk are meant for graduate students in biostatistics.

7

Data Analysis with Intrinsic Estimator and Comparison with Other Methods

Given the properties of the intrinsic estimator, including robust trend estimation with finite samples and consistent estimation with diverging samples, the intrinsic estimator has the potential to analyze APC data and meet the expectation of the investigators. I demonstrate the intrinsic estimator method with the data sets of the lung cancer mortality among the US males, and the HIV mortality data among the US males and females. I then demonstrate the constraint estimator using the lung cancer mortality data among the US females, and further compare the results between the intrinsic estimator and the constrained estimators, using the equality constraints that are commonly used in the literature and the non-contrast constraints that are recently discovered in Fu (2016).

7.1 Illustration of Data Analysis with the Intrinsic Estimator

I apply the intrinsic estimator method to the lung cancer mortality data among the US males and the HIV mortality data among the US males and females. Since the data have mortality rate as well as the population exposure, two models may be fitted, the linear model (4.4) on either the rate or the log-transformed rate, and the loglinear model (7.1) on the mortality frequency with the population exposure.

$$\log(EY_{ij}) = \log(N_{ij}) + \mu + \alpha_i + \beta_j + \gamma_k. \tag{7.1}$$

The APC models use more degrees of freedom than the AP and AC models, and are expected to achieve a better goodness-of-fit. I examine the goodness-of-fit of the above two models, the linear model (4.4) and the loglinear model (7.1).

7.1.1 Modeling Lung Cancer Mortality Data among US Males

7.1.1.1 Intrinsic Estimator of Linear Models

The linear model (4.4) assumes that the mortality rate or the log-transformed rate depends on the additive effects of the age, period and cohort, and the random error ε has mean 0 and common variance σ^2. Table 7.1 provides the output of **R** function `apclinfit` fitted to the log-transformed lung cancer mortality rate among US males. The specification `transform="log"` in the command `apclinfit` is the default and specifies the log-transformation on the response variable. Alternatively, one may use the `"identity"` option to model the original scale of the rate. The assignment `apc=0` specifies the estimation method to be the intrinsic estimator, while other values 1, 2 and 3 specify the method to be constrained estimators on the effects of the age, period and cohort, respectively. The assignment `ModelDiag=T` produces the residual plot against the fitted value for model diagnostics. The model goodness-of-fit has a high R^2 value of 0.99988 and an adjusted R_a^2 value of 0.99979, greater than the AP and AC models with a R^2 of 0.99902 and 0.99983, and an adjusted R_a^2 of 0.99874 and 0.99973, respectively, as shown in Table 3.3 in Chapter 3.

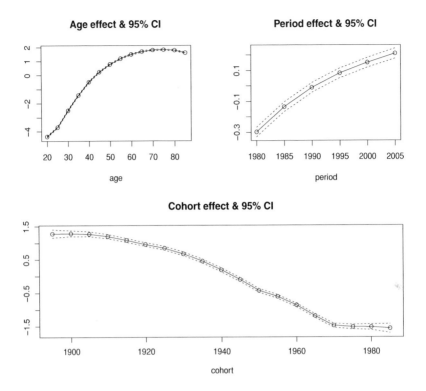

FIGURE 7.1
Plot of age, period and cohort trends by the intrinsic estimator of log-transformed lung cancer mortality rate among US males.

Although the improvement is not dramatic, the key is that it allows simultaneous estimation of the effects of all three factors, the age, period, and cohort.

Figure 7.1 presents the trend in the age, period, and cohort by the parameter estimates of the intrinsic estimator, and the point-wise 95% confidence intervals. It shows an increasing age trend before it plateaus after age 65, an increasing period trend from 1980 to 2005, and an overall decreasing birth cohort trend from the oldest generation born around 1895, to the generation born around 1970, before it flattens through the youngest generations.

TABLE 7.1

Output of Linear APC Model by the Intrinsic Estimator with **R** Function apclinfit on the Lung Cancer Mortality Rate among US Males

```
> apclinfit(r=USlungCAmale.r, apcmodel="APC", header=T,
    transform="log", apc=0, Plot=T)
$model
lm(formula = log(rr) ~ x - 1)
$varcomp
[1]   0.004208291 48.000000000
$F.stats
   value    numdf    dendf   p.value
11338.93    36.00    48.00    0.00
$Rsquared
           R.squared    Adjusted.R.Squared
           0.9998824           0.9997942
$parameter
                 Estimate     Std. Error     t value      Pr(>|t|)
Intercept      3.41201396    0.009281328   367.621295 0.000000e+00
Age 20        -4.35885312    0.029422077  -148.149064 0.000000e+00
Age 25        -3.69014354    0.026694883  -138.234116 0.000000e+00
Age 30        -2.49255913    0.026618982   -93.638408 0.000000e+00
Age 35        -1.40803458    0.026833580   -52.472856 0.000000e+00
Age 40        -0.47193239    0.027199730   -17.350628 0.000000e+00
Age 45         0.23838369    0.027601972     8.636473 2.446265e-11
Age 50         0.78308513    0.027939797    28.027588 0.000000e+00
Age 55         1.19546632    0.027947213    42.775869 0.000000e+00
Age 60         1.49210170    0.027620183    54.022151 0.000000e+00
Age 65         1.69426425    0.027235843    62.207152 0.000000e+00
Age 70         1.80641731    0.026890361    67.177131 0.000000e+00
Age 75         1.82872627    0.026694099    68.506763 0.000000e+00
Age 80         1.78597387    0.026774258    66.704888 0.000000e+00
Age 85         1.59710422    0.028884881    55.292048 0.000000e+00
Period 1980   -0.29805197    0.016052806   -18.566970 0.000000e+00
Period 1985   -0.13428125    0.016126733    -8.326625 7.099143e-11
Period 1990   -0.01040904    0.016145262    -0.644712 5.221833e-01
Period 1995    0.08272514    0.016090955     5.141096 4.977400e-06
Period 2000    0.15069730    0.015966996     9.438050 1.624478e-12
Period 2005    0.20931982    0.016582823    12.622689 0.000000e+00
```

	Estimate	Std. Error	t value	Pr(>|t|)
Cohort 1895	1.28264581	0.062407159	20.552863	0.000000e+00
Cohort 1900	1.28854315	0.044675616	28.842202	0.000000e+00
Cohort 1905	1.27134297	0.037080159	34.286341	0.000000e+00
Cohort 1910	1.20036094	0.032714235	36.692312	0.000000e+00
Cohort 1915	1.08579650	0.029820616	36.410934	0.000000e+00
Cohort 1920	0.96162102	0.027567505	34.882411	0.000000e+00
Cohort 1925	0.84778708	0.028741368	29.497102	0.000000e+00
Cohort 1930	0.67848859	0.029611478	22.913027	0.000000e+00
Cohort 1935	0.45684304	0.030101242	15.176883	0.000000e+00
Cohort 1940	0.19273740	0.030201826	6.381647	6.532349e-08
Cohort 1945	-0.09913810	0.029902041	-3.315429	1.748055e-03
Cohort 1950	-0.42504283	0.029203785	-14.554375	0.000000e+00
Cohort 1955	-0.60750994	0.028109197	-21.612497	0.000000e+00
Cohort 1960	-0.86313037	0.026754689	-32.260900	0.000000e+00
Cohort 1965	-1.18476977	0.028740698	-41.222721	0.000000e+00
Cohort 1970	-1.47794836	0.031551895	-46.841825	0.000000e+00
Cohort 1975	-1.52002567	0.035923505	-42.312844	0.000000e+00
Cohort 1980	-1.52353570	0.043654017	-34.900240	0.000000e+00
Cohort 1985	-1.56506576	0.069332424	-22.573360	0.000000e+00

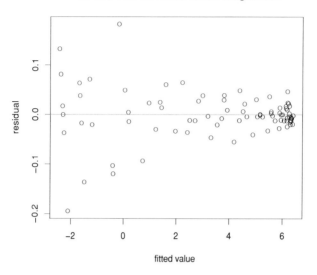

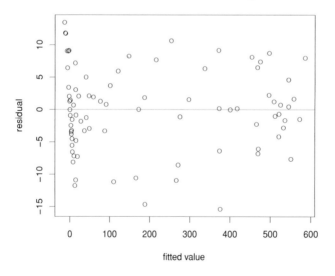

FIGURE 7.2
Residual plots for linear model diagnostics by the intrinsic estimator on lung cancer mortality rate among US males. Upper panel: log-transformed rate; Lower panel: original rate.

The 95% confidence intervals present certain symmetric patterns related to the mortality data structure in the table. The trend and its variability will be compared with the loglinear model of the mortality with the population exposure in the next subsection.

I now examine the residual plot for model diagnostics, a well-known important step in fitting linear models. Figure 7.2 displays two residual plots against the fitted value, one on the log-transformed rate (upper panel) and the other on the original rate with no transformation (lower panel). The log-transformed rate presents decreasing variation of the residuals with the fitted value, indicating inappropriate fit to the log-transformed rate. The decreasing variation further suggests a model on the original rate for better fit. The original rate presents a constant variation around 0, indicating the satisfaction of the Gauss-Markov conditions of the model. Together with large values of $R^2 = 0.99954$ and the adjusted $R_a^2 = 0.99919$ as shown in the model output in Table 7.2, it confirms that the linear model (4.4) fits the lung cancer mortality rates well. Table 7.2 provides the output of **R** function apclinfit, including the model fitting and parameter estimation. Figure 7.3 presents the trends in the age, period, and cohort estimated with the intrinsic estimator method, showing more interesting trends than the log-transformed rate. The age trend shows that the lung cancer mortality rate among US males stays flat and low from age 20 to 45, then starts a sharp increase to a peak around age 80 before it drops till age 85 and older. The period trend shows a steady increase from 1980 to 1990 and then stays flat from 1990 to 2005. The cohort trend shows a sharp increase among the oldest cohorts born from 1895 to 1905 to reach a peak around 1910, followed immediately by a sharp decrease through to the cohort born in 1940, before a mild decline among the young cohorts born from 1940 to 1985. The large cohort effect from 1900 to 1930 and the peak around cohort 1910 may be explained by the large amount of tobacco consumption by these cohorts, as reported in epidemiological studies (Shopland 1995). The flat period trend after 1990 may be attributed to the smoking cessation programs introduced around 1990, which helped to prevent lung cancer.

7.1.1.2   Intrinsic Estimator of Loglinear Models

I fit the loglinear model (7.1) to the lung cancer mortality frequency with the population exposure as the offset. On one hand, the loglinear model may have advantages over linear models. First, it allows estimation of the variance using both pieces of information, including the population exposure and the event frequency. Second, it also allows 0 value of the frequency, with which the calculated 0 mortality rate is not allowed in the linear model on the log-transformed rate, where missing data need to be either omitted or imputed for modeling the log-transformed rate. On the other hand, the loglinear model also has drawbacks, such as the over-dispersion issue, which needs to be examined in the analysis as explained in previous chapters. Another drawback is the estimation of the parameter standard errors, where the PCA method is invalid and the Delta method has to be developed for the computation of the standard errors for the period and cohort effect estimates, as discussed in the previous chapter. I will provide more details of the Delta method in Chapter 9, and compare its performance with the PCA method. For data analysis by the loglinear model in this chapter, I use the Delta method for the standard error estimation, and refer the readers to the theoretical derivation and detailed justification of the Delta method in Chapter 9.

TABLE 7.2

Output of Linear APC Model by the Intrinsic Estimator with **R** Function apclinfit on the Lung Cancer Mortality Rate among US Males

```
> apclinfit(r=USlungCAmale.r, apcmodel="APC", header=T,
     transform="identity", apc=0, Plot=T, ModelDiag = T)
$model
lm(formula = rr ~ x - 1)
$varcomp
[1] 69.79552 48.00000
$F.stats
    value     numdf     dendf   p.value
 2886.668    36.000    48.000     0.000
$Rsquared
             R.squared   Adjusted.R.Squared
             0.9995383            0.9991921
$parameter
                 Estimate     Std. Error      t value      Pr(>|t|)
Intercept      200.338774       1.195284  167.6077148  0.000000e+00
Age 20        -122.765058       3.789084  -32.3996696  0.000000e+00
Age 25        -126.400806       3.437866  -36.7672335  0.000000e+00
Age 30        -130.195685       3.428091  -37.9790654  0.000000e+00
Age 35        -133.553797       3.455727  -38.6470857  0.000000e+00
Age 40        -133.917807       3.502882  -38.2307550  0.000000e+00
Age 45        -126.377310       3.554684  -35.5523344  0.000000e+00
Age 50        -104.586423       3.598190  -29.0663964  0.000000e+00
Age 55         -62.139622       3.599145  -17.2651056  0.000000e+00
Age 60           5.349465       3.557029    1.5039138  1.391554e-01
Age 65          87.866156       3.507532   25.0507037  0.000000e+00
Age 70         169.248533       3.463040   48.8728212  0.000000e+00
Age 75         225.023374       3.437765   65.4563079  0.000000e+00
Age 80         251.405089       3.448088   72.9114532  0.000000e+00
Age 85         201.043891       3.719902   54.0454872  0.000000e+00
Period 1980    -25.402245       2.067340  -12.2874080  2.220446e-16
Period 1985     -5.645662       2.076860   -2.7183638  9.101954e-03
Period 1990      8.253096       2.079246    3.9692726  2.402340e-04
Period 1995      9.247280       2.072253    4.4624287  4.891073e-05
Period 2000      8.339958       2.056289    4.0558303  1.826235e-04
Period 2005      5.207574       2.135597    2.4384625  1.850041e-02
```

	Estimate	Std. Error	t value	Pr(>|t|)
Cohort 1895	24.919580	8.037024	3.1005978	3.228804e-03
Cohort 1900	70.310689	5.753491	12.2205272	2.220446e-16
Cohort 1905	111.935378	4.775320	23.4403946	0.000000e+00
Cohort 1910	125.617887	4.213060	29.8163065	0.000000e+00
Cohort 1915	111.465332	3.840409	29.0243384	0.000000e+00
Cohort 1920	90.755141	3.550245	25.5630642	0.000000e+00
Cohort 1925	75.652394	3.701419	20.4387516	0.000000e+00
Cohort 1930	40.000979	3.813475	10.4893763	5.173639e-14
Cohort 1935	1.284426	3.876549	0.3313324	7.418356e-01
Cohort 1940	-26.587850	3.889503	-6.8357970	1.313053e-08
Cohort 1945	-45.134525	3.850895	-11.7205279	1.110223e-15
Cohort 1950	-55.310333	3.760971	-14.7063965	0.000000e+00
Cohort 1955	-60.066803	3.620006	-16.5930111	0.000000e+00
Cohort 1960	-65.553697	3.445568	-19.0255143	0.000000e+00
Cohort 1965	-73.872593	3.701333	-19.9583741	0.000000e+00
Cohort 1970	-79.171914	4.063370	-19.4843009	0.000000e+00
Cohort 1975	-81.183195	4.626362	-17.5479572	0.000000e+00
Cohort 1980	-82.379608	5.621925	-14.6532728	0.000000e+00
Cohort 1985	-82.681290	8.928885	-9.2599787	2.949863e-12

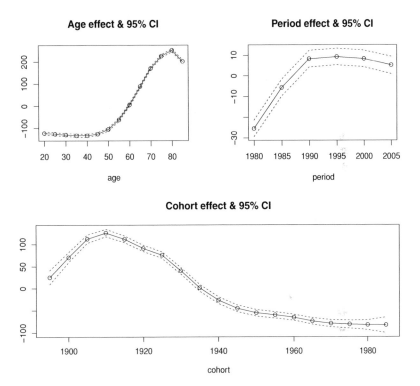

FIGURE 7.3
Plot of age, period and cohort trends by the intrinsic estimator of linear model
(4.4) on lung cancer mortality rate among US males.

Now I apply the loglinear model to the lung cancer mortality data among the US males and compare the results with linear models. Table 7.3 displays the output of the loglinear model (7.1) fitted to the mortality data with **R** function `apcglmfit`. The specification `n.risk=USlungCAmale.p` takes the male population as the model offset term for the population exposure. The model uses the same number of parameters as the linear model. The trend estimates in Figure 7.4 appear to be similar to the ones in the linear model on the log-transformed rate in Figure 7.1. The confidence interval appears to be narrow, and thus over-dispersion may need to be examined.

Examining Over-Dispersion for Loglinear Models   Given the small standard errors and the narrow confidence intervals by the intrinsic estimator, it is necessary to examine the over-dispersion of the loglinear model fitted to the lung cancer mortality data. The model Pearson chi-square statistic is 278.60 on 48 degrees of freedom as shown in Table 7.3, indicating severe over-dispersion in the model. The over-dispersion can be examined by fitting the loglinear model with the quasi-likelihood method (McCullagh and Nelder 1989) using the specification `family="qlik"` in **R** function `apcglmfit`. This allows the APC model to be fitted with the quasi-likelihood approach, adjusting the standard errors of the parameter estimates accordingly.

Table 7.4 displays the output of the loglinear model on the lung cancer mortality data through the quasi-likelihood approach. It is shown that the approach yields the same parameter estimates but with larger standard errors, adjusting for the over-dispersion with a dispersion parameter of 5.80. Figure 7.5 presents age, period, and cohort trends with 95% confidence intervals, showing the same trend estimates but wider confidence intervals, due to the over-dispersion in the loglinear models.

TABLE 7.3
Output of Loglinear APC Model by the Intrinsic Estimator with **R** Function
apcglmfit on the Lung Cancer Mortality Data among US Males

```
> apcglmfit(r=USlungCAmale.r, header=T, n.risk=USlungCAmale.p,
    Scale=1e-5, apcmodel="APC", fam="loglin", apc=0, Plot=T)
$model
glm(formula = rr ~ x - 1 + offset(log(n.risk)),
    family = poisson(link = log))
$deviance
     Deviance degrees of freedom
[1,] 277.3243                    48
$p.val
p.value
      0
$pearson.chisq
[1] 278.6019
$parameter
               Estimate   Std. Error      z value      Pr(>|z|)
Intercept    -8.11098305 0.0089249139 -908.802382 0.000000e+00
Age 20       -4.35991764 0.0555422256  -78.497352 0.000000e+00
Age 25       -3.68300089 0.0342985157 -107.380766 0.000000e+00
Age 30       -2.48108478 0.0203536346 -121.898856 0.000000e+00
Age 35       -1.41191334 0.0144059354  -98.009140 0.000000e+00
Age 40       -0.49049301 0.0112835464  -43.469756 0.000000e+00
Age 45        0.21951963 0.0092175977   23.815275 0.000000e+00
Age 50        0.77307172 0.0075268181  102.708968 0.000000e+00
Age 55        1.18864017 0.0060733509  195.714062 0.000000e+00
Age 60        1.49133295 0.0049562223  300.901143 0.000000e+00
Age 65        1.69547097 0.0044084216  384.598192 0.000000e+00
Age 70        1.81122490 0.0046264458  391.493811 0.000000e+00
Age 75        1.83502319 0.0055233161  332.232152 0.000000e+00
Age 80        1.79487989 0.0068604625  261.626662 0.000000e+00
Age 85        1.61724625 0.0084653063  191.044031 0.000000e+00
Period 1980 -0.27591158 0.0045884507  -60.131752 0.000000e+00
Period 1985 -0.13278598 0.0027340103  -48.568208 0.000000e+00
Period 1990 -0.01289789 0.0008932517  -14.439251 0.000000e+00
Period 1995  0.05637571 0.0009486353   59.428224 0.000000e+00
Period 2000  0.14006679 0.0027535750   50.867251 0.000000e+00
Period 2005  0.22515294 0.0045157961   49.858969 0.000000e+00
```

	Estimate	Std. Error	z value	Pr(>|z|)
Cohort 1895	1.25046113	0.0111385606	112.264157	0.000000e+00
Cohort 1900	1.27564055	0.0096628463	132.014990	0.000000e+00
Cohort 1905	1.26564590	0.0084830651	149.196769	0.000000e+00
Cohort 1910	1.20092553	0.0076017762	157.979594	0.000000e+00
Cohort 1915	1.09093409	0.0071070914	153.499375	0.000000e+00
Cohort 1920	0.96379834	0.0070700780	136.320751	0.000000e+00
Cohort 1925	0.85707828	0.0075153340	114.043937	0.000000e+00
Cohort 1930	0.68766410	0.0083400890	82.452849	0.000000e+00
Cohort 1935	0.46431091	0.0094448946	49.159988	0.000000e+00
Cohort 1940	0.21123472	0.0107406530	19.666842	0.000000e+00
Cohort 1945	-0.08147459	0.0121553989	-6.702749	2.101011e-08
Cohort 1950	-0.38755739	0.0136235922	-28.447518	0.000000e+00
Cohort 1955	-0.57519153	0.0150715178	-38.164141	0.000000e+00
Cohort 1960	-0.80177901	0.0163144459	-49.145341	0.000000e+00
Cohort 1965	-1.17170766	0.0170636347	-68.666945	0.000000e+00
Cohort 1970	-1.55376457	0.0159257188	-97.563230	0.000000e+00
Cohort 1975	-1.59314497	0.0136606239	-116.623149	0.000000e+00
Cohort 1980	-1.53131132	0.0331884000	-46.139956	0.000000e+00
Cohort 1985	-1.57176249	0.1081422413	-14.534214	0.000000e+00

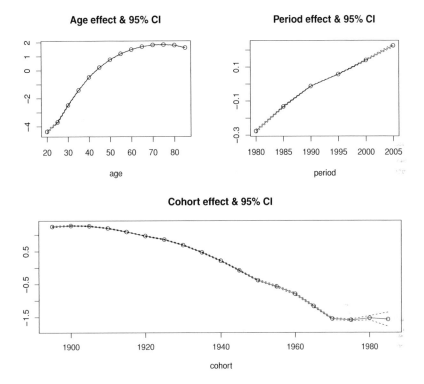

FIGURE 7.4
Plot of age, period and cohort trends by the intrinsic estimator of loglinear model on the lung cancer mortality data among US males.

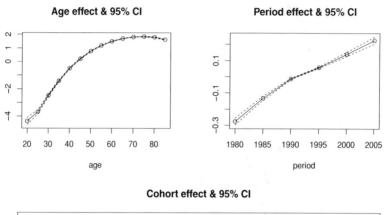

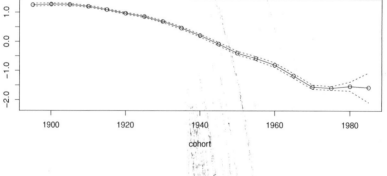

FIGURE 7.5

Plot of age, period, and cohort trends by the intrinsic estimator of the loglinear model with overdispersion adjustment on the lung cancer mortality data among US males.

TABLE 7.4

Output of Loglinear APC Model by the Intrinsic Estimator with the Quasi-Likelihood Approach by **R** Function apcglmfit on the Lung Cancer Mortality Data among US Males

```
> apcglmfit(r=USlungCAmale.r, header=T, n.risk=USlungCAmale.p,
  Scale=1e-5, apcmodel="APC", fam="qlik", apc=0, Plot=T)
$model
glm(formula = rr ~ x - 1 + offset(log(n.risk)),
  family = quasipoisson(link = log))
$deviance
    Deviance degrees of freedom
[1,] 277.3243                 48
$p.val
p.value
      0
$pearson.chisq
[1] 278.6019
$dispersion
[1] 5.804206
$parameter
```

	Estimate	Std. Error	t value	Pr(>|t|)
Intercept	-8.11098305	0.021501830	-377.222909	0.000000e+00
Age 20	-4.35991764	0.133811881	-32.582440	0.000000e+00
Age 25	-3.68300089	0.082631707	-44.571279	0.000000e+00
Age 30	-2.48108478	0.049035812	-50.597404	0.000000e+00
Age 35	-1.41191334	0.034706663	-40.681334	0.000000e+00
Age 40	-0.49049301	0.027184229	-18.043293	0.000000e+00
Age 45	0.21951963	0.022206962	9.885172	3.690381e-13
Age 50	0.77307172	0.018133549	42.632124	0.000000e+00
Age 55	1.18864017	0.014631868	81.236393	0.000000e+00
Age 60	1.49133295	0.011940491	124.897125	0.000000e+00
Age 65	1.69547097	0.010620734	159.637840	0.000000e+00
Age 70	1.81122490	0.011145996	162.500052	0.000000e+00
Age 75	1.83502319	0.013306728	137.901904	0.000000e+00
Age 80	1.79487989	0.016528171	108.595194	0.000000e+00
Age 85	1.61724625	0.020394548	79.297971	0.000000e+00
Period 1980	-0.27591158	0.011054458	-24.959303	0.000000e+00
Period 1985	-0.13278598	0.006586756	-20.159543	0.000000e+00
Period 1990	-0.01289789	0.002152015	-5.993400	2.566189e-07
Period 1995	0.05637571	0.002285445	24.667285	0.000000e+00
Period 2000	0.14006679	0.006633891	21.113823	0.000000e+00
Period 2005	0.22515294	0.010879419	20.695308	0.000000e+00

	Estimate	Std. Error	t value	Pr(>|t|)
Cohort 1895	1.25046113	0.026834930	46.598263	0.000000e+00
Cohort 1900	1.27564055	0.023279651	54.796378	0.000000e+00
Cohort 1905	1.26564590	0.020437332	61.928138	0.000000e+00
Cohort 1910	1.20092553	0.018314138	65.573686	0.000000e+00
Cohort 1915	1.09093409	0.017122347	63.714050	0.000000e+00
Cohort 1920	0.96379834	0.017033175	56.583600	0.000000e+00
Cohort 1925	0.85707828	0.018105882	47.337008	0.000000e+00
Cohort 1930	0.68766410	0.020092875	34.224276	0.000000e+00
Cohort 1935	0.46431091	0.022754564	20.405177	0.000000e+00
Cohort 1940	0.21123472	0.025876294	8.163252	1.249241e-10
Cohort 1945	-0.08147459	0.029284689	-2.782157	7.696495e-03
Cohort 1950	-0.38755739	0.032821848	-11.807909	8.881784e-16
Cohort 1955	-0.57519153	0.036310179	-15.841055	0.000000e+00
Cohort 1960	-0.80177901	0.039304631	-20.399098	0.000000e+00
Cohort 1965	-1.17170766	0.041109571	-28.502065	0.000000e+00
Cohort 1970	-1.55376457	0.038368113	-40.496247	0.000000e+00
Cohort 1975	-1.59314497	0.032911065	-48.407579	0.000000e+00
Cohort 1980	-1.53131132	0.079957225	-19.151632	0.000000e+00
Cohort 1985	-1.57176249	0.260535413	-6.032817	2.234033e-07

Now I consider the model goodness-of-fit, by examining the model deviance and the degrees of freedom. In general, a model deviance close to the degrees of freedom indicates a good fit. The large model deviance of 277.32 relative to the 48 degrees of freedom indicates a poor fit to the data by the loglinear model. Recall that the linear model on the log-transformed rate also presents an improper fit to the data as shown in the residual plot in Figure 7.2. Hence, it may suggest additive age, period, and cohort effects as in the linear model on the original rate, but not multiplicative as in the loglinear model or the linear model on the log-transformed rate, although it may seem to be reasonable to assume that the number of lung cancer deaths follows a Poisson distribution.

Therefore in comparing the results of the linear models with those of the loglinear models, it would be more appropriate to conclude that a linear model on the original scale of the lung cancer mortality rate among US males achieves a proper fit to the data. The corresponding age, period, and cohort trends by the estimates present sensible interpretation of the lung cancer mortality among US males from the epidemiological point of view.

7.1.2 Modeling the HIV Mortality Data

I now analyze the HIV mortality data using the linear model (4.4) on the mortality rate and the loglinear model (7.1) on the frequency and the population exposure.

7.1.2.1 Intrinsic Estimator of Linear Models

First, I fit the linear model to the mortality rate data on the original scale. Table 7.5 displays the output of the linear model (4.4). Although it is shown that the model achieves large values of $R^2 = 0.939$ and the adjusted $R_a^2 = 0.893$, and age, period, and cohort effects present somewhat reasonable trends

TABLE 7.5

Output of Linear APC Model by the Intrinsic Estimator with **R** Function apclinfit on the HIV Mortality Rate among US Males and Females

```
> UShiv.r = apcheader(UShiv.r, agestart=20, agespan=5,
  yearstart=1987, yearspan=5, header=F)
> apclinfit(r=UShiv.r, apcmodel = "APC", header=T, transform="id",
  apc=0, Plot=T,  ModelDiag=T)
$model
lm(formula = rr ~ x - 1)
$transform
[1] "id"
$varcomp
[1] 12.09813 48.00000
$F.stats
   value    numdf    dendf   p.value
20.40347 36.00000 48.00000  0.00000
$Rsquared
          R.squared   Adjusted.R.Squared
          0.9386601            0.8926552
$parameter
             Estimate Standard.Error          t       P-value
Intercept    5.07727410     0.4976413 10.202678869 1.307843e-13
Age 20      -1.05017813     1.5775371 -0.665707410 5.087848e-01
Age 25       3.10008536     1.4313119  2.165904873 3.531642e-02
Age 30       7.86967915     1.4272423  5.513905556 1.373237e-06
Age 35       8.94836020     1.4387485  6.219544562 1.157365e-07
Age 40       6.82857545     1.4583805  4.682300388 2.356879e-05
Age 45       2.11622252     1.4799477  1.429930599 1.592141e-01
Age 50      -2.24569538     1.4980610 -1.499068034 1.404047e-01
Age 55      -4.82058342     1.4984586 -3.217028032 2.321407e-03
Age 60      -5.79955631     1.4809241 -3.916173857 2.838975e-04
Age 65      -5.54692788     1.4603168 -3.798441473 4.097284e-04
Age 70      -4.70533726     1.4417929 -3.263531954 2.031250e-03
Age 75      -3.29819957     1.4312698 -2.304387010 2.557166e-02
Age 80      -1.40813753     1.4355678 -0.980892421 3.315641e-01
Age 85       0.01169281     1.5487340  0.007549916 9.940074e-01
Period 1987  2.74416560     0.8607107  3.188255401 2.520088e-03
Period 1992  8.06092366     0.8646745  9.322495189 2.391420e-12
Period 1997 -1.03290548     0.8656680 -1.193188970 2.386598e-01
Period 2002 -1.88600398     0.8627562 -2.186021958 3.372315e-02
Period 2007 -3.42098154     0.8561098 -3.995961196 2.208155e-04
Period 2012 -4.46519826     0.8891289 -5.021992011 7.479552e-06
```

	Estimate	Standard.Error	t	P-value
Cohort 1903	-7.43313250	3.3461135	-2.221422718	3.107288e-02
Cohort 1908	-9.33159636	2.3953931	-3.895642952	3.027579e-04
Cohort 1913	-6.23645392	1.9881440	-3.136822041	2.915782e-03
Cohort 1918	-3.94882366	1.7540543	-2.251255127	2.898440e-02
Cohort 1923	-1.78093186	1.5989058	-1.113844162	2.708926e-01
Cohort 1928	0.29713686	1.4780996	0.201026273	8.415275e-01
Cohort 1933	2.48584957	1.5410392	1.613099475	1.132790e-01
Cohort 1938	4.60877587	1.5876923	2.902814274	5.571494e-03
Cohort 1943	6.30637219	1.6139522	3.907409512	2.918056e-04
Cohort 1948	7.90072007	1.6193453	4.878959608	1.215946e-05
Cohort 1953	9.73483873	1.6032716	6.071858957	1.947279e-07
Cohort 1958	8.52329948	1.5658328	5.443301111	1.754828e-06
Cohort 1963	4.73652135	1.5071438	3.142713586	2.867655e-03
Cohort 1968	-1.26273152	1.4345185	-0.880247608	3.831113e-01
Cohort 1973	-5.56774538	1.5410033	-3.613065198	7.227859e-04
Cohort 1978	-4.91798843	1.6917325	-2.907072070	5.507574e-03
Cohort 1983	-3.79307497	1.9261272	-1.969275463	5.470834e-02
Cohort 1988	-1.25913782	2.3406176	-0.537951098	5.930965e-01
Cohort 1993	0.93810229	3.7174286	0.252352469	8.018461e-01

in Figure 7.6, the downward-then-upward age trend from age 40 to 85 years old may not be supported. I then examine the model diagnostics by the residual plot against the fitted value. The deviation of the residuals from 0, as shown in Figure 7.7, indicates a severe violation of the Gauss-Markov conditions of linear models. This suggests that the linear model does not constitute a good fit to the original scale of the HIV mortality rate.

Second, I fit the linear model to the log-transformed mortality rate. Table 7.6 displays the output of the linear model (4.4) on the log-transformed rate. The model achieves large values of $R^2 = 0.997$ and adjusted $R_a^2 = 0.995$, and age, period, and cohort effects present reasonable trends, as shown in Figure 7.8. It is shown that it has a sharply increasing age trend from age 20 to the peak around age 35 followed by a sharp decrease through age 85. The period estimates present a very interesting trend of sharp changes between

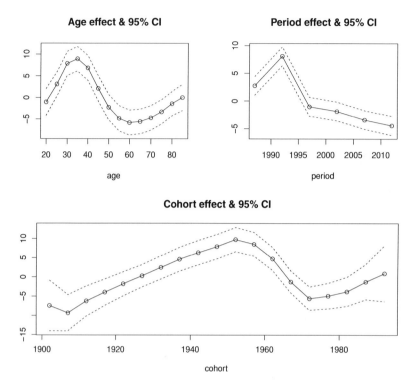

FIGURE 7.6
Plot of age, period, and cohort trends by the intrinsic estimator of the linear
model fitted to the HIV mortality rate among US males and females.

three pieces of linear pattern, a sharp increase from 1987 to the peak around
1992, followed by a steep decrease to 1997, before a steady decline till 2012 but
at a much slower pace. The cohort effects present a more interesting trend, a
fast increase from the old cohorts born between 1903 and 1913 to the peak
around the cohort 1953, followed by a fast decline to the young cohorts born
between 1983 and 1993. The wider confidence intervals at the extreme old and
young cohorts are due to the small number of observations on these cohorts.
I examine the model diagnostics by the residual plot against the fitted value.

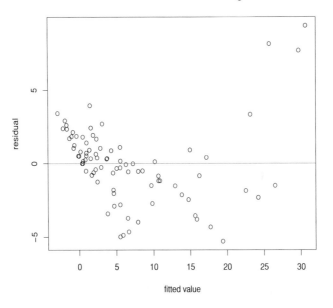

FIGURE 7.7

Residual plot of linear model fitted to the HIV mortality rate among US males and females.

As shown in Figure 7.9, there is no major deviation of the residuals from 0 except for one outlier that has a relatively large value greater than 0.3. The variation of the residuals seems to remain constant with the fitted value. The residual plot indicates that the Gauss-Markov conditions are satisfied by the linear model on the log-transformed rate. It can thus be concluded that the linear model (4.4) achieved a good fit to the log-transformed HIV mortality rate among the US males and females.

TABLE 7.6
Output of Linear APC Model by the Intrinsic Estimator with **R** Function apclinfit on the Log-Transformed HIV Mortality Rate among the US Males and Females

```
> apclinfit(r=UShiv.r, apcmodel="APC", header=T, transform="log",
  apc=0, Plot=T, ModelDiag=T)
$model
lm(formula = log(rr) ~ x - 1)
$transform
[1] "log"
$varcomp
[1]   0.01390415 48.00000000
$F.stats
   value    numdf    dendf   p.value
503.2762  36.0000  48.0000   0.0000
$Rsquared
           R.squared    Adjusted.R.Squared
           0.9973577             0.9953760
$parameter
```

	Estimate	Standard.Error	t	P-value
Intercept	0.915831628	0.01687056	54.2857986	0.000000e+00
Age 20	-0.550829268	0.05348015	-10.2996962	9.547918e-14
Age 25	0.483262272	0.04852296	9.9594557	2.893241e-13
Age 30	0.897156640	0.04838500	18.5420419	0.000000e+00
Age 35	0.975688127	0.04877507	20.0038289	0.000000e+00
Age 40	0.907238067	0.04944061	18.3500567	0.000000e+00
Age 45	0.724848681	0.05017177	14.4473424	0.000000e+00
Age 50	0.474687803	0.05078583	9.3468561	2.204015e-12
Age 55	0.219143994	0.05079931	4.3139171	7.955682e-05
Age 60	-0.043117659	0.05020487	-0.8588342	3.947007e-01
Age 65	-0.285351752	0.04950626	-5.7639532	5.739438e-07
Age 70	-0.542473453	0.04887828	-11.0984566	7.549517e-15
Age 75	-0.778170305	0.04852153	-16.0376277	0.000000e+00
Age 80	-1.090148557	0.04866724	-22.4000493	0.000000e+00
Age 85	-1.391934590	0.05250369	-26.5111753	0.000000e+00
Period 1987	0.391845436	0.02917899	13.4290276	0.000000e+00
Period 1992	0.693955639	0.02931336	23.6736953	0.000000e+00
Period 1997	-0.099159347	0.02934704	-3.3788529	1.452677e-03
Period 2002	-0.165138627	0.02924833	-5.6460870	8.665132e-07
Period 2007	-0.329611834	0.02902301	-11.3569134	3.330669e-15
Period 2012	-0.491891268	0.03014239	-16.3189183	0.000000e+00

	Estimate	Standard.Error	t	P-value
Cohort 1903	-0.832033206	0.11343673	-7.3347781	2.259680e-09
Cohort 1908	-1.022409549	0.08120632	-12.5902711	0.000000e+00
Cohort 1913	-0.960609590	0.06740015	-14.2523365	0.000000e+00
Cohort 1918	-0.635000495	0.05946426	-10.6786911	2.819966e-14
Cohort 1923	-0.257899182	0.05420457	-4.7578865	1.829219e-05
Cohort 1928	0.085996149	0.05010911	1.7161778	9.257777e-02
Cohort 1933	0.378056508	0.05224283	7.2365241	3.193465e-09
Cohort 1938	0.592659049	0.05382442	11.0109697	9.769963e-15
Cohort 1943	0.717284275	0.05471466	13.1095452	0.000000e+00
Cohort 1948	0.857960288	0.05489749	15.6284072	0.000000e+00
Cohort 1953	0.997383504	0.05435257	18.3502540	0.000000e+00
Cohort 1958	0.972919213	0.05308336	18.3281392	0.000000e+00
Cohort 1963	0.833483756	0.05109374	16.3128344	0.000000e+00
Cohort 1968	0.507034141	0.04863167	10.4260072	6.350476e-14
Cohort 1973	0.009461487	0.05224161	0.1811102	8.570437e-01
Cohort 1978	-0.391190806	0.05735149	-6.8209351	1.383829e-08
Cohort 1983	-0.651288657	0.06529771	-9.9741419	2.757794e-13
Cohort 1988	-0.635548613	0.07934937	-8.0094979	2.130545e-10
Cohort 1993	-0.566258272	0.12602469	-4.4932327	4.418545e-05

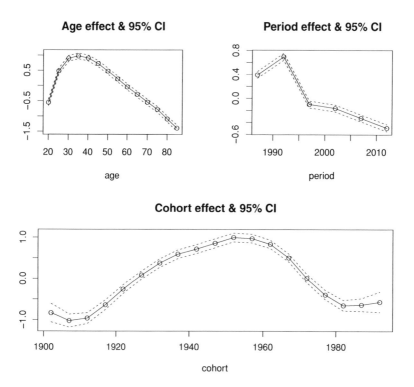

FIGURE 7.8
Plot of age, period, and cohort trends by the intrinsic estimator of linear
model fitted to the log-transformed HIV mortality rate among the US males
and females.

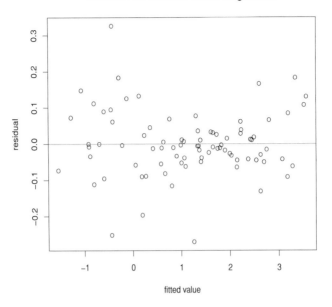

FIGURE 7.9
Residual plot of linear model fitted to the log-transformed HIV mortality rate among US males and females.

7.1.2.2 Intrinsic Estimator of Loglinear Models

I now fit the loglinear model (7.1) to the US HIV mortality data. Table 7.7 displays the output of **R** function apcglmfit by the intrinsic estimator of the loglinear model by the maximum likelihood approach. The standard errors of the period and cohort effect estimates are computed with the Delta method. The model goodness-of-fit seems to be poor with a large deviance 925.9 on 48 degrees of freedom. The overall age, period and cohort trends are similar to the linear model (7.1) on the log-transformed rate. The confidence intervals seem to be too narrow in Figure 7.10. An examination of the model Pearson chi-square of 929.1 on 48 degrees of freedom indicates severe over-dispersion.

I then fit the loglinear model with the quasi-likelihood approach by specifying the option `family = "qlik"` in **R** function `apcglmfit`. Table 7.8 displays the output of **R** function taking the quasi-likelihood approach. It yields the same model deviance and degrees of freedom, and presents severe overdispersion with a large dispersion parameter of 19.4. It is also shown that the standard errors of age, period, and cohort effect estimates are larger than those by the maximum likelihood approach. Figure 7.11 presents age, period, and cohort trends by the parameter estimates and the 95% confidence intervals. It is shown that the confidence intervals by the quasi-likelihood approach are much wider and seem to be reasonable.

The loglinear model presents age, period, and cohort trends similar to those by the linear model on the log-transformed rate, as shown in Figure 7.11. The age effect presents a sharp increasing-then-decreasing trend with a peak around age 30–35. The period trend has three pieces of linear pattern, a sharp increase from 1987 to 1992, a sharp decrease from 1992 to 1997, and a steady decline from 1997 to 2012 at a much slower pace than the previous period from 1992 to 1997. The cohort has a sharp increasing trend from the old cohorts born between 1903 and 1913 to a peak around the cohort 1953, followed by a sharp decline to the young cohorts born between 1983 and 1993.

Overall, comparing the output of the loglinear model by the quasi-likelihood approach with the linear model on the log-transformed rate, I conclude that the loglinear model (7.1) with the quasi-likelihood approach achieves a good fit to the HIV mortality data, and the linear model (4.4) also achieves a good fit to the log-transformed HIV mortality rate among US males and females.

TABLE 7.7

Output of Loglinear APC Model by the Intrinsic Estimator with **R** Function apcglmfit on the HIV Mortality Data among US Males and Females

```
> apcglmfit(r=UShiv.r, header=T, n.risk=UShiv.p, apcmodel="APC",
  family="loglin", Plot=T, apc=0, n.interval = 5)
$model
glm(formula = rr ~ x - 1 + offset(log(n.risk)),
  family = poisson(link = log))
$deviance
    Deviance degrees of freedom
[1,] 925.9386                 48
$pearson.chisq
[1] 929.1093
$p.val
p.value
      0
$parameter.delta
```

	Estimate	Std. Error	z value	Pr(>|z|)
Intercept	-10.590070835	0.009644081	-1098.090202	0.000000e+00
Age 20	-0.656064842	0.016944688	-38.718025	0.000000e+00
Age 25	0.485745562	0.012309727	39.460302	0.000000e+00
Age 30	0.940090096	0.009801222	95.915598	0.000000e+00
Age 35	1.001542157	0.007922495	126.417516	0.000000e+00
Age 40	0.920150944	0.006837490	134.574384	0.000000e+00
Age 45	0.731272006	0.006933428	105.470483	0.000000e+00
Age 50	0.486997384	0.008173306	59.583890	0.000000e+00
Age 55	0.232874410	0.010183222	22.868441	0.000000e+00
Age 60	-0.025014119	0.012670186	-1.974250	5.412252e-02
Age 65	-0.274420682	0.015636249	-17.550288	0.000000e+00
Age 70	-0.538769568	0.019648704	-27.420107	0.000000e+00
Age 75	-0.802066895	0.025339055	-31.653386	0.000000e+00
Age 80	-1.109670455	0.035327041	-31.411362	0.000000e+00
Age 85	-1.392666000	0.049456901	-28.159184	0.000000e+00
Period 1987	0.325669277	0.006530536	49.868693	0.000000e+00
Period 1992	0.807888422	0.003885310	207.934112	0.000000e+00
Period 1997	-0.118865608	0.001319063	-90.113648	0.000000e+00
Period 2002	-0.154143738	0.001419319	-108.603991	0.000000e+00
Period 2007	-0.361447986	0.003997379	-90.421249	0.000000e+00
Period 2012	-0.499100368	0.006335399	-78.779625	0.000000e+00

	Estimate	Std. Error	z value	Pr(>|z|)
Cohort 1903	-0.767140922	0.081476862	-9.415445	1.751932e-12
Cohort 1908	-1.032741193	0.025647909	-40.266097	0.000000e+00
Cohort 1913	-0.857119879	0.017518795	-48.925733	0.000000e+00
Cohort 1918	-0.652853215	0.019817672	-32.942983	0.000000e+00
Cohort 1923	-0.296203640	0.020553846	-14.411105	0.000000e+00
Cohort 1928	0.050187676	0.020064834	2.501275	1.583930e-02
Cohort 1933	0.329277473	0.018702281	17.606274	0.000000e+00
Cohort 1938	0.544715384	0.016765477	32.490300	0.000000e+00
Cohort 1943	0.696155731	0.014587734	47.721991	0.000000e+00
Cohort 1948	0.833760739	0.012312451	67.716877	0.000000e+00
Cohort 1953	0.972637304	0.010070421	96.583579	0.000000e+00
Cohort 1958	0.938961282	0.008037675	116.820016	0.000000e+00
Cohort 1963	0.816790772	0.006487110	125.909809	0.000000e+00
Cohort 1968	0.518844907	0.005728841	90.567165	0.000000e+00
Cohort 1973	-0.009550581	0.006255878	-1.526657	1.334093e-01
Cohort 1978	-0.417186976	0.008084717	-51.601926	0.000000e+00
Cohort 1983	-0.639928683	0.010492478	-60.989278	0.000000e+00
Cohort 1988	-0.566584267	0.014595419	-38.819322	0.000000e+00
Cohort 1993	-0.462021911	0.033333429	-13.860618	0.000000e+00

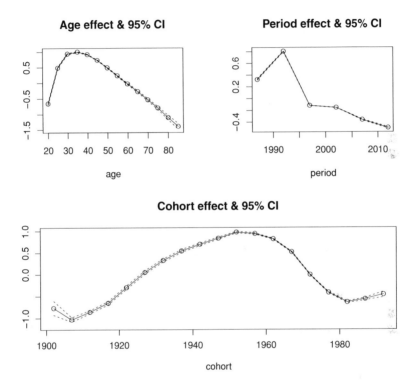

FIGURE 7.10

Plot of age, period, and cohort trends by the intrinsic estimator of the loglinear model fitted to the HIV mortality data among US males and females.

TABLE 7.8

Output of Loglinear APC Model by the Intrinsic Estimator with Quasi-Likelihood Approach by **R** Function `apcglmfit` on the Mortality Data among US Males and Females

```
> apcglmfit(r=UShiv.r, header=T, n.risk=UShiv.p, apcmodel="APC",
  fam="qlik", Plot=T, apc=0)
$model
glm(formula = rr ~ x - 1 + offset(log(n.risk)),
  family = quasipoisson(link = log))
$deviance
    Deviance degrees of freedom
[1,] 925.9386                48
$pearson.chisq
[1] 929.1093
$p.val
p.value
      0
$dispersion
[1] 19.35645
$parameter.delta
```

	Estimate	Std. Error	t value	Pr(>|t|)
Intercept	-10.590070835	0.042430061	-249.5888644	0.000000e+00
Age 20	-0.656064842	0.074549783	-8.8003589	1.397504e-11
Age 25	0.485745562	0.054157829	8.9690737	7.875034e-12
Age 30	0.940090096	0.043121419	21.8010007	0.000000e+00
Age 35	1.001542157	0.034855779	28.7338911	0.000000e+00
Age 40	0.920150944	0.030082193	30.5878949	0.000000e+00
Age 45	0.731272006	0.030504283	23.9727647	0.000000e+00
Age 50	0.486997384	0.035959247	13.5430362	0.000000e+00
Age 55	0.232874410	0.044802064	5.1978500	4.096245e-06
Age 60	-0.025014119	0.055743699	-0.4487345	6.556426e-01
Age 65	-0.274420682	0.068793182	-3.9890680	2.256802e-04
Age 70	-0.538769568	0.086446361	-6.2324146	1.106016e-07
Age 75	-0.802066895	0.111481608	-7.1946118	3.701595e-09
Age 80	-1.109670455	0.155424711	-7.1396012	4.493625e-09
Age 85	-1.392666000	0.217590391	-6.4004021	6.113795e-08
Period 1987	0.325669277	0.028731719	11.3348343	3.552714e-15
Period 1992	0.807888422	0.017093793	47.2620909	0.000000e+00
Period 1997	-0.118865608	0.005803346	-20.4822547	0.000000e+00
Period 2002	-0.154143738	0.006244432	-24.6849909	0.000000e+00
Period 2007	-0.361447986	0.017586852	-20.5521702	0.000000e+00
Period 2012	-0.499100368	0.027873197	-17.9061038	0.000000e+00

	Estimate	Std. Error	t value	Pr(>|t|)
Cohort 1903	-0.767140922	0.358465286	-2.1400703	3.745946e-02
Cohort 1908	-1.032741193	0.112840441	-9.1522258	4.239276e-12
Cohort 1913	-0.857119879	0.077075621	-11.1205056	7.105427e-15
Cohort 1918	-0.652853215	0.087189751	-7.4877288	1.320009e-09
Cohort 1923	-0.296203640	0.090428621	-3.2755519	1.962034e-03
Cohort 1928	0.050187676	0.088277165	0.5685239	5.723297e-01
Cohort 1933	0.329277473	0.082282482	4.0017932	2.167791e-04
Cohort 1938	0.544715384	0.073761328	7.3848370	1.894885e-09
Cohort 1943	0.696155731	0.064180140	10.8469027	1.643130e-14
Cohort 1948	0.833760739	0.054169812	15.3916122	0.000000e+00
Cohort 1953	0.972637304	0.044305785	21.9528285	0.000000e+00
Cohort 1958	0.938961282	0.035362522	26.5524408	0.000000e+00
Cohort 1963	0.816790772	0.028540664	28.6184926	0.000000e+00
Cohort 1968	0.518844907	0.025204588	20.5853360	0.000000e+00
Cohort 1973	-0.009550581	0.027523335	-0.3469994	7.301085e-01
Cohort 1978	-0.417186976	0.035569490	-11.7287871	1.110223e-15
Cohort 1983	-0.639928683	0.046162667	-13.8624720	0.000000e+00
Cohort 1988	-0.566584267	0.064213948	-8.8233831	1.292078e-11
Cohort 1993	-0.462021911	0.146653624	-3.1504295	2.805754e-03

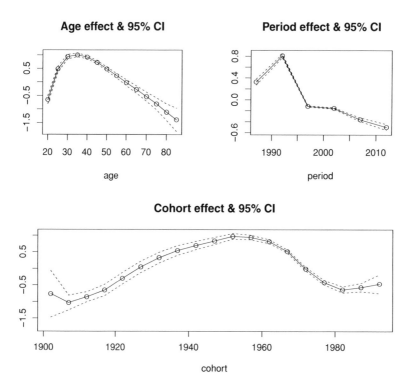

FIGURE 7.11
Plot of age, period, and cohort trends by the intrinsic estimator of loglinear model by the quasi-likelihood approach on the HIV mortality data among US males and females.

7.2 Illustration of Data Analysis with Constrained Estimators

Now I examine the constrained estimators: the equality constraints and the non-contrast constraints. The former are popular but yield biased estimation, while the latter have been only discovered recently and yield consistent estimation. Although both linear model (4.4) and loglinear model (7.1) may be fitted with a pre-selected constraint, I here examine the loglinear model only on the lung cancer mortality data among the US females. Similar results may be obtained with the linear model, on the lung cancer mortality data among the US males, and on the HIV mortality data among US males and females. I leave them as exercises for the readers.

7.2.1 Illustration of Equality Constraints

I now illustrate the constrained estimators by specifying an equality constraint on age, period, or cohort effects using **R** function `apcglmfit` with the specification `apc=1` for constraints on age effects, `apc=2` for constraints on period effects, and `apc=3` for constraints on cohort effects. Details of the constraints are further specified by selecting the effects of the constraint. For example, the option `lindex=c(1,2)` specifies the first two effects are involved, and their coefficients are specified by `lvalue=c(-1,1)`, indicating the coefficients of the selected effects in the constraints are $(-1, 1)$. Hence the constraint is $-\alpha_1 + \alpha_2 = 0$, or equivalently $\alpha_1 = \alpha_2$, if the age effect constraint is specified by `apc=1`. Table 7.9 displays the output from a constrained estimator with a constraint assuming identical first two age effects. It has the same number of parameters and thus the same degrees of freedom as the intrinsic estimator method. Figure 7.12 displays age, period, and cohort trends estimated with

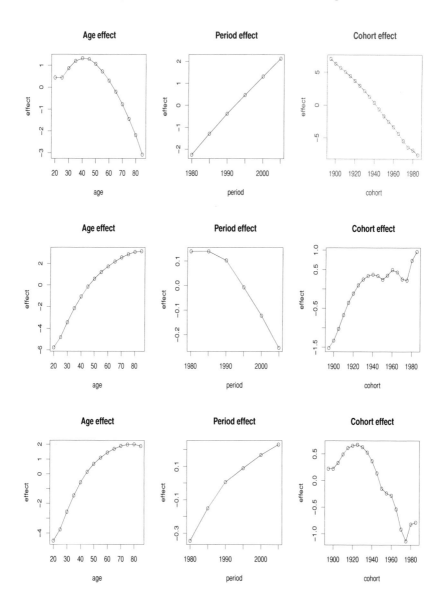

FIGURE 7.12

Plot of age, period, and cohort trends estimated with equality constraints on lung cancer mortality data among US females. Top panels: assuming identical effects of the two youngest age groups, or $\alpha_{20} = \alpha_{25}$; Middle panels: assuming identical effects of the two earliest periods, or $\beta_{1980} = \beta_{1985}$; Bottom panels: assuming identical effects of the two oldest cohorts, or $\gamma_{1895} = \gamma_{1900}$.

TABLE 7.9
Output of Loglinear APC Model on the Lung Cancer Mortality Rate among US Females by **R** Function `apcglmfit` with Equality Constraint $\alpha_{20} = \alpha_{25}$

```
> apcglmfit(r=USlungCAfemale.r, header=T,
  n.risk=USlungCAfemale.p, Scale=1e-5,
  apcmodel="APC", fam="loglin", apc=1,
  lindex=c(1,2), lvalue=c(-1,1), Plot=T)
$model
glm(formula = rr ~ x + offset(log(n.risk)),
  family = poisson(link = log))
$constraint
[1] "Age"
$constraint.para
       [,1] [,2]
lindex    1    2
lvalue   -1    1
$deviance
[1] 583.6165  48.0000
$p.val
p.value
      0
$parameter
               Estimate
Intercept    -8.7401265
Age 20        0.4382571
Age 25        0.4382571
Age 30        0.8642119
Age 35        1.2040185
Age 40        1.3311641
Age 45        1.2903206
Age 50        1.0670702
Age 55        0.7246906
Age 60        0.3047104
Age 65       -0.2008352
Age 70       -0.7707895
Age 75       -1.4377809
Age 80       -2.1807859
Age 85       -3.0725089
```

```
                Estimate
Period 1980 -2.2486382
Period 1985 -1.2934805
Period 1990 -0.3750896
Period 1995  0.4706876
Period 2000  1.3113089
Period 2005  2.1352119
Cohort 1895  7.0766811
Cohort 1900  6.3152000
Cohort 1905  5.6631118
Cohort 1910  5.0601289
Cohort 1915  4.4199908
Cohort 1920  3.7014387
Cohort 1925  2.9579034
Cohort 1930  2.1522770
Cohort 1935  1.2887559
Cohort 1940  0.3663044
Cohort 1945 -0.6224928
Cohort 1950 -1.6741547
Cohort 1955 -2.5237106
Cohort 1960 -3.3293437
Cohort 1965 -4.3463034
Cohort 1970 -5.4841470
Cohort 1975 -6.4678865
Cohort 1980 -6.9139654
Cohort 1985 -7.6397877
```

an equality constraint: identical first two age effects in the top panels, identical first two period effects in the middle panels, and identical first two cohort effects in the bottom panels. It is clearly shown that all three panels display largely different trend estimates, demonstrating the estimation bias by the equality constraint method, which has been reported in the literature.

7.2.2   Illustration of Non-Contrast Constraints

I now examine the non-contrast constraints. Two kinds of non-contrast constraints are studied, one on period effects and the other on cohort effects. Both have been proven to yield consistent estimation of the age effects. I illustrate the non-contrast constraint method using the lung cancer mortality data among US females, and compare the results with the intrinsic estimator.

Table 7.10 displays the output by the non-contrast constraint $\beta_1 = 2\beta_2$. The estimates are close to the intrinsic estimator, as shown in Figure 7.13. However, this does not imply that one may impose an arbitrary non-contrast constraint on the period or cohort effects because the consistency only holds true on age effect estimates but not necessarily true on period or cohort effects. One needs to use caution when selecting a non-contrast constraint in order to yield reasonable estimates close to the consistent ones, such as the intrinsic estimator.

Figure 7.14 presents the estimated trends by two non-contrast constraints; $\gamma_{1895} = 2\gamma_{1900}$ in the upper panels and $\gamma_{1930} = 2\gamma_{1940}$ in the lower panels. Although both non-contrast constraints yield consistent estimation of the age effects, and present robust age trend estimates, as shown in the plot, the period and cohort trends differ largely. By the consistency and the fast convergence of the intrinsic estimator to the true parameters, the constraint $\gamma_{1895} = \gamma_{1900}$ yields trends different from the intrinsic estimator and thus is not reliable. In contrast, the constraint $\gamma_{1930} = 2\gamma_{1940}$ yields estimates close to the intrinsic estimator and thus is reliable. It presents reasonable trend estimation, an increasing age trend, increasing period trend, and an overall decreasing cohort trend.

In general, non-contrast constraints may provide good parameter estimation due to the consistency of the age trend, but may also present biased estimation for the period and cohort effects if an inappropriate constraint is specified. Caution should be used in specifying such constraints. Readers may find it useful to select a constraint on the period or cohort effects by examining the effect estimates of the intrinsic estimator.

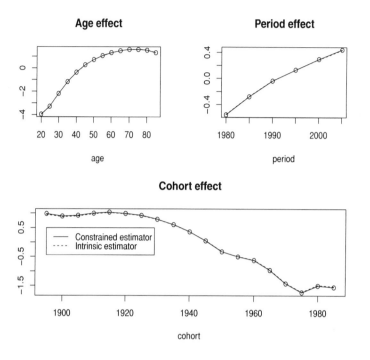

FIGURE 7.13
Plot of estimated trend by non-contrast constraint $\beta_{1980} = 2\beta_{1985}$ compared to the intrinsic estimator on the lung cancer mortality rate among US females.

The non-contrast constraints provide an alternative method to specify a constraint that may lead to consistent estimation without prior knowledge of the event under investigation. By the symmetry between the rows and columns of the APC data, non-contrast constraints on the cohort effects are recommended, which may potentially lead to consistent estimation on both the age and period effects, depending on which dimension of the APC data diverging to the infinity, the number of age groups or that of the periods. Further studies are needed for the behavior of the period effect estimates when the non-contrast constraints on the cohort effects are specified.

TABLE 7.10
Output of Loglinear APC Model by **R** Function `apcglmfit` with Non-Contrast
Constraint $\beta_{1980} = 2\beta_{1985}$ on Lung Cancer Mortality Data among US Females

```
> apcglmfit(r=USlungCAfemale.r, header=T,
  n.risk=USlungCAfemale.p, Scale=1e-5, apcmodel="APC",
  fam="loglin", apc=2, lindex=c(1,2), lvalue=c(-1,2), Plot=T)


$model
glm(formula = rr ~ x + offset(log(n.risk)),
  family = poisson(link = log))
$constraint
[1] "Period"
$constraint.para
       [,1] [,2]
lindex    1    2
lvalue    1   -2
$deviance
[1] 583.6165  48.0000
$p.val
p.value
      0
$parameter
              Estimate
Intercept   -8.74012653
Age 20      -3.95993782
Age 25      -3.28329244
Age 30      -2.18069230
Age 35      -1.16424032
Age 40      -0.36044935
Age 45       0.27535250
Age 50       0.72874747
Age 55       1.06301328
Age 60       1.31967844
Age 65       1.49077827
Age 70       1.59746930
Age 75       1.60712332
Age 80       1.54076366
Age 85       1.32568599
```

```
                 Estimate
Period 1980 -0.55702479
Period 1985 -0.27851239
Period 1990 -0.03676696
Period 1995  0.13236490
Period 2000  0.29634079
Period 2005  0.44359845
Cohort 1895  0.98687269
Cohort 1900  0.90203704
Cohort 1905  0.92659415
Cohort 1910  1.00025666
Cohort 1915  1.03676392
Cohort 1920  0.99485716
Cohort 1925  0.92796725
Cohort 1930  0.79898628
Cohort 1935  0.61211048
Cohort 1940  0.36630443
Cohort 1945  0.05415259
Cohort 1950 -0.32086399
Cohort 1955 -0.49377451
Cohort 1960 -0.62276221
Cohort 1965 -0.96307653
Cohort 1970 -1.42427478
Cohort 1975 -1.73136887
Cohort 1980 -1.50080244
Cohort 1985 -1.54997931
```

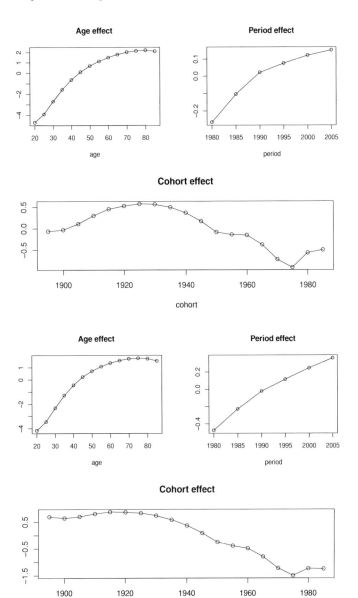

FIGURE 7.14
Plot of estimated trend by non-contrast constraints on the lung cancer mortality rate among US females. Upper panels: $\gamma_{1895} = 2\gamma_{1900}$; Lower panels: $\gamma_{1930} = 2\gamma_{1940}$.

7.3 Suggested Readings

The equality constraint method has been studied in many papers, including the early work by Rodgers (1982), Kupper et al (1985), Holford (1985), Clayton and Schifflers (1987), Robertson and Boyle (1998). The non-contrast constraint method was discovered in Fu (2016), while the intrinsic estimator method was developed as a limit of the ridge estimator in Fu (2000), and has been studied in a number of papers, including Fu (2000), Yang, Fu and Land (2004), Fu (2008), Yang et al (2008), and Fu (2016).

7.4 Exercises

1. Fit a loglinear APC model to the lung cancer mortality data among US females. Examine the model deviance and over-dispersion.

2. Fit a linear APC model to the lung cancer mortality rate among US females, on both the original scale and the log scale. What happens to the 0 mortality rate in the age group 20–24? Drop the missing data and fit the model again. Examine the residual plot for the model diagnostics. What can you conclude from the model fitting?

3.* Why is the age trend of the HIV mortality rate in Figure 7.7 not supported? In particular, can you explain from the behavioral, epidemiological and public health points of view why the upward trend from age 60 to 85 following a downtrend between age 35 and 60 does not make sense?

---

* Difficult exercises with an asterisk are meant for graduate students in biostatistics.

4.* Can you explain why the age trend for the loglinear model by the quasi-likelihood approach has a wide confidence interval for the oldest age groups? Does the same answer apply to the wide confidence intervals for the oldest and the youngest cohorts in the same figure?

Part II

A Resolution to the Identification Problem: Theoretical Justification and Related Topics

8

Asymptotic Behavior of Multiple Estimators — Theoretical Results

In Chapter 6, I provided the rationale and briefly explained why the intrinsic estimator presents reasonable results, and why the popular equality constraints hardly yield reasonable results even though the specification of the constraints is often based on reasonable assumptions about the event under investigation. In this chapter, I provide rigorous theoretical results to address these issues. These results help to achieve a thorough understanding of the multiple estimators and various methods in the APC analysis, especially for the graduate students in statistics and biostatistics and those statisticians who are interested in understanding the complex identification problem. In particular, these results help to explain why the identification problem can be resolved even though multiple estimators exist for a given data set. This approach requires advanced graduate level statistics, in particular, statistical large sample theory. Readers who are not familiar with asymptotic theory and the related techniques may skip this chapter to move on to other interesting topics, including the variance estimation.

8.1 Settings and Strategies to Study the Asymptotics of Multiple Estimators

Asymptotic studies of estimators have been well established in statistics. Often the estimators are unique for given models. On one hand, the age-period-cohort models yield multiple estimators, presenting major challenges to the asymptotic studies. On the other hand, if the behavior of each estimator is known, it may potentially provide a clue to understanding and further resolving the identification problem. From both theoretical and practical points of view, it is always desirable to analyze data using statistical methods of good properties, such as unbiased estimates and efficient estimation with high accuracy, rather than using an arbitrary method with unknown behavior. This motivates the asymptotic studies of the multiple estimators.

To study the asymptotics, one needs to focus on one estimator at a time. Hence, a penalty approach is taken to focus on the asymptotics of one specific estimator first, and extend the result later to other estimators. Given that there are infinitely many estimators, the structure of estimators (6.1) provides an efficient approach to extending the results from one estimator to others with ease.

8.2 Assumptions and Regularity Conditions for the Asymptotics

I explained in Chapter 6 why the asymptotic studies in APC analysis take the limit as $p \to \infty$. For the asymptotics, I assume the response variable follows a distribution in the exponential family. Specifically, assume the response Y_{ij}

with $i = 1, \ldots, a$ and $j = 1, \ldots, p$ follows an exponential family distribution with log-likelihood (McCullagh and Nelder 1989)

$$l(\boldsymbol{\zeta}, \boldsymbol{y}) = [\boldsymbol{y}^T \boldsymbol{\zeta} - \psi(\boldsymbol{\zeta})]/\kappa(\phi) + c(\boldsymbol{y}, \phi), \tag{8.1}$$

where $\kappa(\phi)$ is a dispersion parameter, $\boldsymbol{\zeta}$ is a parameter vector with link function $g(\boldsymbol{\zeta}) = \boldsymbol{\eta}$ for model $\boldsymbol{\eta} = X\boldsymbol{b}$. Denote $\boldsymbol{b}^T = (\boldsymbol{\theta}^T, \boldsymbol{\xi}^T)$, where $\boldsymbol{\theta}^T = (\mu, \alpha_1, \ldots, \alpha_{a-1})$ are the parameters of primary interest and $\boldsymbol{\xi}^T = (\beta_1, \ldots, \beta_{p-1}, \gamma_1, \ldots, \gamma_{a+p-2})$ are the parameters of secondary interest. I derive the consistency of the estimator for $\boldsymbol{\theta}$ as $p \to \infty$. Notice that maximizing the log-likelihood (8.1) yields multiple estimators due to the singularity of matrix X. I thus first consider the intrinsic estimator $\boldsymbol{B}$ using the following penalized log-likelihood, and then consider the multiple estimators $\boldsymbol{B} + t\boldsymbol{B}_0$ for $t \neq 0$. Let

$$l_p^\lambda(\boldsymbol{b}; \boldsymbol{y}) = \sum_{i=1}^{a} \sum_{j=1}^{p} l\left(\boldsymbol{\zeta}(\boldsymbol{b}); y_{ij}\right) - \lambda(\boldsymbol{b}^T \boldsymbol{B}_0)^2, \tag{8.2}$$

where $\lambda > 0$ is the tuning parameter for the penalty term $(\boldsymbol{b}^T \boldsymbol{B}_0)^2$. The parameter λ need not be selected since $l_p^\lambda(\boldsymbol{b}; \boldsymbol{y})$ is maximized at $\boldsymbol{b} = \boldsymbol{B}$ with penalty $(\boldsymbol{B}^T \boldsymbol{B}_0)^2 = 0$. For the same reason, an infinitesimal λ is preferred, i.e. $\lambda << 1$ as long as it is positive. This will make it convenient to prove the consistency and to calculate the variance later. Furthermore, the matrix X is dropped from the likelihood function $l(\boldsymbol{b}, y_{ij})$ for the simplicity of notation since the design matrix X remains the same. I study the consistency of the estimator of $\boldsymbol{\theta}$ through the penalized profile log-likelihood

$$Pl_p^\lambda(\boldsymbol{\theta}; \boldsymbol{y}) = l_p^\lambda(\boldsymbol{\theta}, \boldsymbol{\xi}_{\boldsymbol{\theta}}^\lambda; \boldsymbol{y}), \tag{8.3}$$

where $\boldsymbol{\xi}_{\boldsymbol{\theta}}^\lambda = \arg\max_{\boldsymbol{\xi}} l_p^\lambda(\boldsymbol{\theta}, \boldsymbol{\xi}; \boldsymbol{y})$ for each fixed $\boldsymbol{\theta}$ and $\lambda > 0$. Such $\boldsymbol{\xi}_{\boldsymbol{\theta}}^\lambda$ exists because l_p^λ is concave in $(\boldsymbol{\theta}, \boldsymbol{\xi})$ for the exponential family distributions. The parameters $\boldsymbol{\xi}_{\boldsymbol{\theta}}^\lambda$ are profile functions of $\boldsymbol{\theta}$. I study the maximum profile likelihood estimator (MaPLE) by maximizing the profile log-likelihood

$$\tilde{\boldsymbol{\theta}} = \arg\max_{\boldsymbol{\theta}} Pl_p^\lambda(\boldsymbol{\theta}; \boldsymbol{y}).\tag{8.4}$$

For the consistency of the estimator $\tilde{\boldsymbol{\theta}}$, I put it in the framework of profile like-lihood to make the derivation of the asymptotic results technically tractable.

Remark 8.1 The parameter $\lambda > 0$ can be infinitesimal to serve the purpose of attaining the intrinsic estimator $\boldsymbol{B}$ to be the unique MaPLE as long as λ is positive.

For the key consistency results, the following regularity conditions on the likelihood function are needed.

The Regularity Conditions

1) The parameter space $\mathbb{D} = D_1 \times D_2 \times \cdots \times D_{2a+2p-3} \subset \mathbf{R}^{2a+2p-3}$ for model parameters $\boldsymbol{b}$ has its sub-spaces D_j uniformly bounded with respect to p for $j = 1, \ldots, 2a + 2p - 3$.

2) The log-likelihood $l(\boldsymbol{b}; \boldsymbol{y})$ of the exponential family is continuously differentiable with derivatives $\partial^2 l / \partial b_i \partial b_j$ and a uniform bound $|\partial l / \partial b_j| < M < \infty$ for $\boldsymbol{b} \in \mathbb{D}$, $i, j = 1, \ldots, 2a + 2p - 3$.

Regularity condition 1 implies uniform boundedness of age, period, and cohort effects with respect to p, but not the compactness of $\mathbb{D}$ because the dimension increases with p. Regularity condition 2 ensures the smoothness needed for the profile likelihood.

8.3   Asymptotics of Multiple Estimators

I now give several lemmas to prepare for the major asymptotic results.

Lemma 1 *Assume that the log-likelihood* $l(\boldsymbol{\theta}, \boldsymbol{\xi}; \boldsymbol{y})$ *satisfies regularity conditions 1 and 2. Then*

1. *for given $\lambda > 0$, the profile $\boldsymbol{\xi}_{\boldsymbol{\theta}}^{\lambda} = \boldsymbol{\xi}^{\lambda}(\boldsymbol{\theta})$ is continuously differentiable with respect to $\boldsymbol{\theta}$.*

2. *the functions $\boldsymbol{\beta}_j = \boldsymbol{\xi}_{cj}^{\lambda}(\boldsymbol{\theta})$ for j-th period and $\gamma_k = \boldsymbol{\xi}_{dk}^{\lambda}(\boldsymbol{\theta})$ for k-th cohort are continuously differentiable and the derivatives are bounded for $j = 1, \ldots, L_1$ and $k = 1, \ldots, L_2$ with fixed $L_1 < p$ and fixed $L_2 < a + p - 1$.*

3. *The profile log-likelihood $Pl_p^{\lambda}(\boldsymbol{\theta}; \boldsymbol{y})$ leads to a sub-model and the MaPLE $\tilde{\boldsymbol{\theta}}$ is independent of λ.*

Proof   *It is straightforward to see that for any given $\lambda > 0$, the penalized log-likelihood function $l_p^{\lambda}(\boldsymbol{\theta}, \boldsymbol{\xi}; y)$ is concave and continuously differentiable with respect to $(\boldsymbol{\theta}, \boldsymbol{\xi})$ for the exponential family distributions. By the definition of the profile, $\boldsymbol{\xi}_{\boldsymbol{\theta}}^{\lambda}$ maximizes the log-likelihood $l_p^{\lambda}(\boldsymbol{\theta}, \boldsymbol{\xi}; \boldsymbol{y})$ for each fixed $\boldsymbol{\theta}$, $\partial l_p^{\lambda}(\boldsymbol{\theta}, \boldsymbol{\xi}; \boldsymbol{y})/\partial \boldsymbol{\xi}|_{\boldsymbol{\xi} = \boldsymbol{\xi}_{\boldsymbol{\theta}}^{\lambda}} = 0$. By the chain rule*

$$\frac{\partial \boldsymbol{\xi}_{\boldsymbol{\theta}}^{\lambda}}{\partial \boldsymbol{\theta}} = \left(-\frac{\partial^2 l_p^{\lambda}(\boldsymbol{\theta}, \boldsymbol{\xi}; \boldsymbol{y})}{\partial \boldsymbol{\xi}^2} \right)^{-1} \frac{\partial^2 l_p^{\lambda}(\boldsymbol{\theta}, \boldsymbol{\xi}; \boldsymbol{y})}{\partial \boldsymbol{\xi} \partial \boldsymbol{\theta}}.$$

Let $(\boldsymbol{\theta}_0, \boldsymbol{\xi}_0)$ denote the partition of the non-random null vector $\boldsymbol{B}_0$, i.e., $\boldsymbol{B}_0^T = (\boldsymbol{\theta}_0^T, \boldsymbol{\xi}_0^T)$.

$$-\frac{\partial^2 l_p^{\lambda}(\boldsymbol{\theta}, \boldsymbol{\xi}; \boldsymbol{y})}{\partial \boldsymbol{\xi}^2} = -\frac{\partial^2 l(\boldsymbol{\theta}, \boldsymbol{\xi}; \boldsymbol{y})}{\partial \boldsymbol{\xi}^2} + 2\lambda \boldsymbol{\xi}_0 \otimes \boldsymbol{\xi}_0$$

is positive-definite and bounded in the parameter space $\mathbb{D}$ under the regularity conditions, where $\otimes$ is the outer product.

$$\frac{\partial^2 l_p^{\lambda}(\boldsymbol{\theta}, \boldsymbol{\xi}; \boldsymbol{y})}{\partial \boldsymbol{\xi} \partial \boldsymbol{\theta}} = \frac{\partial^2 l(\boldsymbol{\theta}, \boldsymbol{\xi}; \boldsymbol{y})}{\partial \boldsymbol{\xi} \partial \boldsymbol{\theta}} + 2\lambda \boldsymbol{\xi}_0 \otimes \boldsymbol{\theta}_0$$

is also bounded in $\mathbb{D}$. The derivative $(\partial \boldsymbol{\xi}_{\boldsymbol{\theta}}^{\lambda}/\partial \boldsymbol{\theta})$ is thus continuous and bounded in $\mathbb{D}$ and, in particular, is a constant in the special case of a Gaussian response. Hence, $\boldsymbol{\xi}_{\boldsymbol{\theta}}^{\lambda}$ is a function of $\boldsymbol{\theta}$, and leads to a sub-model (Murphy and van der Vaart 2000) under the profile log-likelihood $Pl_p^{\lambda}(\boldsymbol{\theta}; \boldsymbol{y})$. Since the estimator $\widehat{\boldsymbol{b}} = \boldsymbol{B}$ satisfies $\boldsymbol{b}^T \boldsymbol{B}_0 = \boldsymbol{0}$, the penalized log-likelihood (8.2) has a maximizer $\boldsymbol{B} = (\widehat{\boldsymbol{\theta}}^T, \widehat{\boldsymbol{\xi}}^T)^T$ independent of the parameter λ. The MaPLE $\tilde{\boldsymbol{\theta}}$ under the profile log-likelihood $Pl_p^\lambda(\boldsymbol{\theta}; \boldsymbol{y})$ is numerically equal to $\hat{\boldsymbol{\theta}}$, so $\tilde{\boldsymbol{\theta}}$ and its profile $\boldsymbol{\xi}_{\tilde{\boldsymbol{\theta}}}$ are independent of λ. ■

Since $\tilde{\boldsymbol{\theta}}$ and its profile $\boldsymbol{\xi}_{\tilde{\boldsymbol{\theta}}}$ are independent of λ, λ is dropped from the notation hereafter.

Lemma 2 *The profile log-likelihood $Pl_p^\lambda(\boldsymbol{\theta}; \boldsymbol{y})$ of the MaPLE $\tilde{\boldsymbol{\theta}}_p$ converges in probability*

$$\frac{Pl_p^\lambda(\tilde{\boldsymbol{\theta}}_p)}{p} \longrightarrow_p -\frac{a}{2}\kappa(\phi) \qquad \text{as} \qquad p \to \infty.$$

Proof   *For the simplicity of notation, I write the model deviance $Dev(\boldsymbol{b}; \boldsymbol{y}) = 2l_p^\lambda(\boldsymbol{y}; \boldsymbol{y}) - 2l_p^\lambda(\boldsymbol{b}; \boldsymbol{y})$ (McCullagh and Nelder 1989, p.24) as $Dev(\boldsymbol{b}; \boldsymbol{y}) = -2l_p^\lambda(\boldsymbol{b}; \boldsymbol{y})$ by absorbing the first term into the log-likelihood. The deviance is asymptotically distributed as $\kappa(\phi)\chi_d^2$ with d degrees of freedom and dispersion parameter $\kappa(\phi)$. The sub-model has a parameters under the profile log-likelihood; thus the degrees of freedom $d = (ap - a)$. Hence, the sub-model deviance $Dev(\boldsymbol{\theta}; \boldsymbol{y})$ satisfies*

$$\frac{Dev(\boldsymbol{\theta}; \boldsymbol{y})}{p} = \frac{Dev(\boldsymbol{\theta}, \boldsymbol{\xi}_{\boldsymbol{\theta}}^\lambda; \boldsymbol{y})}{p} \sim \frac{\kappa(\phi)\chi_{ap-a}^2}{p} \quad \text{as} \quad p \to \infty,$$

following the discussion in McCullagh and Nelder (1989, p.118). Let Z_p be a random variable following χ^2 distribution with $(ap - a)$ degrees of freedom, i.e. $Z_p \sim \chi_{ap-a}^2$. By Chebyshev's inequality (Williams 1991, p.73) on the random variable $Z_p\kappa(\phi)/p$, for any fixed $\varepsilon > 0$,

$$P\left(\left|\frac{\kappa(\phi)Z_p}{p} - E[\frac{\kappa(\phi)Z_p}{p}]\right| > \varepsilon\right) \leq var\left[\frac{\kappa(\phi)Z_p}{p}\right]/\varepsilon^2.$$

Since

$$E[\frac{\kappa(\phi)Z_p}{p}] = \kappa(\phi)(ap - a)/p \to a\kappa(\phi) \text{ as } p \to \infty, \text{ and}$$

$$var[\frac{\kappa(\phi)Z_p}{p}] = 2(ap - a)\kappa^2(\phi)/p^2,$$

$$P\left(\left|\frac{\kappa(\phi)Z_p}{p} - E[\frac{\kappa(\phi)Z_p}{p}]\right| > \varepsilon\right) \le 2(ap-a)\kappa^2(\phi)/(p^2\varepsilon^2) < 2a\kappa^2(\phi)/(p\varepsilon^2).$$

The right-hand side converges to 0 as $p \to \infty$. It is then followed by the convergence in probability $\kappa(\phi)Z_p/p \to_p a\kappa(\phi)$ as $p \to \infty$. Hence,

$$\frac{Pl_p^\lambda(\tilde{\boldsymbol{\theta}}_p)}{p} \to_p -\frac{a}{2}\kappa(\phi).$$

∎

Lemma 3 *For any $\lambda > 0$, the partial derivative of the profile log-likelihood satisfies the following convergence in probability,*

$$-\frac{1}{p}\frac{\partial^2 Pl_p^\lambda(\boldsymbol{\theta};\boldsymbol{y})}{\partial\boldsymbol{\theta}^2} \longrightarrow_p C_1 \qquad \text{as} \qquad p \to \infty, \tag{8.5}$$

where C_1 is an $a \times a$ Fisher information matrix of the age effect model.

Proof *By the profile likelihood, for any $\boldsymbol{\theta}$,*

$$\partial l_p^\lambda(\boldsymbol{\theta},\boldsymbol{\xi};\boldsymbol{y})/\partial\boldsymbol{\xi}|_{\boldsymbol{\xi}=\boldsymbol{\xi}_{\boldsymbol{\theta}}} = 0.$$

Furthermore,

$$(\partial^2 l_p^\lambda(\boldsymbol{\theta},\boldsymbol{\xi};\boldsymbol{y})/\partial\boldsymbol{\xi}\partial\boldsymbol{\theta})|_{\boldsymbol{\xi}=\boldsymbol{\xi}_{\boldsymbol{\theta}}} = 0$$

for smooth likelihood functions l under the regularity conditions. Thus for large p,

$$-\frac{\partial^2 Pl_p^\lambda(\boldsymbol{\theta};\boldsymbol{y})}{\partial\boldsymbol{\theta}^2} = -\frac{\partial^2 l_p^\lambda(\boldsymbol{\theta},\boldsymbol{\xi};\boldsymbol{y})}{\partial\boldsymbol{\theta}^2} \approx pC_1 + 2\lambda\boldsymbol{\theta}_0 \otimes \boldsymbol{\theta}_0,$$

where C_1 is the Fisher information matrix for the age effect model $l(\boldsymbol{\theta};\boldsymbol{y})$ with parameters $\boldsymbol{\theta}$, which is positive-definite and independent of λ. The conclusion of Lemma 3 follows the fact that the elements of vector $\boldsymbol{\theta}_0$ are of order $O(p^{-3/2})$ because the norm of the null vector $\boldsymbol{v}$ in (6.2) has the order $\|\boldsymbol{v}\| = O(p^{3/2})$. ∎

8.3.1 Asymptotics of Multiple Estimators with Fixed t

I consider the asymptotics of the intrinsic estimator and multiple estimators with a fixed t.

Theorem 8.1 *Under the regularity conditions, the MaPLE $\tilde{\boldsymbol{\theta}}_p$ converges in probability to a unique constant vector $\boldsymbol{\theta}^{\infty}$*

$$\tilde{\boldsymbol{\theta}}_p \to_p \boldsymbol{\theta}^{\infty} \qquad as \quad p \to \infty.$$

Proof *Consider a Cauchy sequence $\{\tilde{\boldsymbol{\theta}}_{p+n}\}$ of the MaPLE $\tilde{\boldsymbol{\theta}}_p$ for an arbitrary large number p and any finite number n. Take the Taylor expansion of the profile log-likelihood $Pl_p^{\lambda}(\boldsymbol{\theta}; \boldsymbol{y})$ with observations $\boldsymbol{y}$ in a table of a rows and an increasing number $(p+n)$ of columns, where the log-likelihood $l_p^{\lambda}(\boldsymbol{\theta}, \boldsymbol{\xi}; \boldsymbol{y})$ assumes the values for $\boldsymbol{y}$ in the first p columns.*

$$Pl_p^{\lambda}(\tilde{\boldsymbol{\theta}}_{p+n}; \boldsymbol{y}) = Pl_p^{\lambda}(\tilde{\boldsymbol{\theta}}_p; \boldsymbol{y}) + \left.\frac{\partial Pl_p^{\lambda}(\boldsymbol{\theta}; \boldsymbol{y})}{\partial \boldsymbol{\theta}}\right|_{\boldsymbol{\theta}=\tilde{\boldsymbol{\theta}}_p} (\tilde{\boldsymbol{\theta}}_{p+n} - \tilde{\boldsymbol{\theta}}_p)$$

$$+ \frac{1}{2}(\tilde{\boldsymbol{\theta}}_{p+n} - \tilde{\boldsymbol{\theta}}_p)^T \left.\frac{\partial^2 Pl_p^{\lambda}(\boldsymbol{\theta}; \boldsymbol{y})}{\partial \boldsymbol{\theta}^2}\right|_{\boldsymbol{\theta}=\tilde{\boldsymbol{\theta}}_p} (\tilde{\boldsymbol{\theta}}_{p+n} - \tilde{\boldsymbol{\theta}}_p) + o(\|\tilde{\boldsymbol{\theta}}_{p+n} - \tilde{\boldsymbol{\theta}}_p\|^2)$$

$$= \quad Pl_p^{\lambda}(\tilde{\boldsymbol{\theta}}_p; \boldsymbol{y}) + \frac{1}{2}(\tilde{\boldsymbol{\theta}}_{p+n} - \tilde{\boldsymbol{\theta}}_p)^T \left.\frac{\partial^2 Pl_p^{\lambda}(\boldsymbol{\theta}; \boldsymbol{y})}{\partial \boldsymbol{\theta}^2}\right|_{\boldsymbol{\theta}=\tilde{\boldsymbol{\theta}}_p} (\tilde{\boldsymbol{\theta}}_{p+n} - \tilde{\boldsymbol{\theta}}_p)$$

$$+ o(\|\tilde{\boldsymbol{\theta}}_{p+n} - \tilde{\boldsymbol{\theta}}_p\|^2), \tag{8.6}$$

where the partial derivative $\partial Pl_p^{\lambda}(\boldsymbol{\theta}; \boldsymbol{y})/\partial \boldsymbol{\theta}|_{\boldsymbol{\theta}=\tilde{\boldsymbol{\theta}}_p} = 0$ because $\tilde{\boldsymbol{\theta}}_p$ maximizes $Pl_p^{\lambda}(\boldsymbol{\theta}; \boldsymbol{y})$.

Consider the log-likelihood function $l(\boldsymbol{\theta}, \boldsymbol{\xi}; \boldsymbol{y})$ on the $a \times (p+n)$ table and its three values at the MaPLEs and the profiles: $l_{p+n}(\tilde{\boldsymbol{\theta}}_{p+n}, \boldsymbol{\xi}_{\tilde{\boldsymbol{\theta}}_{p+n}}; \boldsymbol{y})$, $l_p(\tilde{\boldsymbol{\theta}}_{p+n}, \boldsymbol{\xi}_{\tilde{\boldsymbol{\theta}}_{p+n}}; \boldsymbol{y})$ and $l_p(\tilde{\boldsymbol{\theta}}_p, \boldsymbol{\xi}_{\tilde{\boldsymbol{\theta}}_p}; \boldsymbol{y})$. The first one maximizes the profile likelihood on the $a \times (p+n)$ table. The third one sets the probabilities to 1 in the last n columns, i.e. columns $(p+1)$ to $(p+n)$, and maximizes the profile likelihood on the $a \times p$ sub-table formed by the first p columns. Note that the penalized log-likelihood is equal to the log-likelihood in equation (8.2) at the MaPLE because the penalty term is equal to 0. The second takes the first log-likelihood $l_{p+n}(\tilde{\boldsymbol{\theta}}_{p+n}, \boldsymbol{\xi}_{\tilde{\boldsymbol{\theta}}_{p+n}}; \boldsymbol{y})$ and sets the probabilities to 1 on the last n

columns $(p+1), \ldots, (p+n)$. *Hence its log-likelihood is between the other two,*
i.e. $l_{p+n}(\tilde{\boldsymbol{\theta}}_{p+n}, \boldsymbol{\xi}_{\tilde{\boldsymbol{\theta}}_{p+n}}; \boldsymbol{y}) \leq l_p(\tilde{\boldsymbol{\theta}}_{p+n}, \boldsymbol{\xi}_{\tilde{\boldsymbol{\theta}}_{p+n}}; \boldsymbol{y}) \leq l_p(\tilde{\boldsymbol{\theta}}_p, \boldsymbol{\xi}_{\tilde{\boldsymbol{\theta}}_p}; \boldsymbol{y})$. *Furthermore,*
it is straightforward to show that the penalty term $(\boldsymbol{b}^T \boldsymbol{B}_0)^2$ *of the penalized*
log-likelihood $l_p^\lambda(\tilde{\boldsymbol{\theta}}_{p+n}, \boldsymbol{\xi}_{\tilde{\boldsymbol{\theta}}_{p+n}}; \boldsymbol{y})$ *has an upper bound of the order* $O(p)$, *inde-*
pendent of λ, *uniformly in the parameter space* $\mathbb{D}$.

For large p, $\forall \lambda > 0$, $n > 0$,

$$\frac{1}{p}|Pl_{p+n}^\lambda(\tilde{\boldsymbol{\theta}}_{p+n}; \boldsymbol{y}) - Pl_p^\lambda(\tilde{\boldsymbol{\theta}}_{p+n}; \boldsymbol{y})|$$

$$\leq \frac{1}{p}|l_{p+n}(\tilde{\boldsymbol{\theta}}_{p+n}, \boldsymbol{\xi}_{\tilde{\boldsymbol{\theta}}_{p+n}}; \boldsymbol{y}) - l_p(\tilde{\boldsymbol{\theta}}_{p+n}, \boldsymbol{\xi}_{\tilde{\boldsymbol{\theta}}_{p+n}}; \boldsymbol{y})| + \frac{\lambda}{p}((\tilde{\boldsymbol{\theta}}_{p+n}^T, \boldsymbol{\xi}_{\tilde{\boldsymbol{\theta}}_{p+n}}^T) \boldsymbol{B}_0)^2$$

$$\leq \frac{1}{p}|l_{p+n}(\tilde{\boldsymbol{\theta}}_{p+n}, \boldsymbol{\xi}_{\tilde{\boldsymbol{\theta}}_{p+n}}; \boldsymbol{y}) - l_p(\tilde{\boldsymbol{\theta}}_p, \boldsymbol{\xi}_{\tilde{\boldsymbol{\theta}}_p}; \boldsymbol{y})| + \frac{\lambda}{p} O(p)$$

$$= \left| \frac{p+n}{p} \frac{1}{p+n} Pl_{p+n}^\lambda(\tilde{\boldsymbol{\theta}}_{p+n}; \boldsymbol{y}) - \frac{1}{p} Pl_p^\lambda(\tilde{\boldsymbol{\theta}}_p; \boldsymbol{y}) \right| + \lambda O(1). \qquad (8.7)$$

By Lemma 2, the first term on the right-hand side of (8.7) converges to 0
in probability for any given n *as* $p \to \infty$. *The second term converges to 0 as*
$\lambda \to 0+$. *By equation (8.6) and Lemma 3,*

$$(\tilde{\boldsymbol{\theta}}_{p+n} - \tilde{\boldsymbol{\theta}}_p)^T [C_1 + o(1)](\tilde{\boldsymbol{\theta}}_{p+n} - \tilde{\boldsymbol{\theta}}_p) \leq \frac{2}{p} \left| Pl_p^\lambda(\tilde{\boldsymbol{\theta}}_{p+n}; \boldsymbol{y}) - Pl_p^\lambda(\tilde{\boldsymbol{\theta}}_p; \boldsymbol{y}) \right|$$

$$\leq \frac{2}{p} \left| Pl_p^\lambda(\tilde{\boldsymbol{\theta}}_{p+n}; \boldsymbol{y}) - Pl_{p+n}^\lambda(\tilde{\boldsymbol{\theta}}_{p+n}; \boldsymbol{y}) \right|$$

$$+ \frac{2}{p} \left| \frac{p+n}{p} \frac{1}{p+n} Pl_{p+n}^\lambda(\tilde{\boldsymbol{\theta}}_{p+n}; \boldsymbol{y}) - Pl_p^\lambda(\tilde{\boldsymbol{\theta}}_p; \boldsymbol{y}) \right| \to_p 0, \qquad (8.8)$$

which implies the convergence in probability of the Cauchy sequence $\{\tilde{\boldsymbol{\theta}}_{p+n}\}$
as $p \to \infty$ *and* $\lambda \to 0+$ *since the MaPLEs* $\tilde{\boldsymbol{\theta}}_{p+n}$ *and* $\tilde{\boldsymbol{\theta}}_p$ *are independent of*
λ. *Thus there exists a unique a-dimensional non-random vector* $\boldsymbol{\theta}^\infty$ *such that*
$\tilde{\boldsymbol{\theta}}_p \to_p \boldsymbol{\theta}^\infty$. ∎

Remark 8.2 As discussed in Chapter 6, the unique limit of the intrinsic
estimator serves as the true parameters based on the rationale of the con-
straint approach. I refer to the vector $\boldsymbol{\theta}^\infty$ as true parameters. Consequently,

the intrinsic estimator is consistent. I will give full justification later in this chapter.

Corollary 1 *For a fixed $t \neq 0$, an estimator $B + tB_0$ yields consistent estimation of the model intercept and age effects with a fixed number of age groups.*

Proof *As mentioned previously in the proof of Lemma 3, the norm $\|v\| = O(p^{3/2})$. Thus $\boldsymbol{\theta}_0 = A_0 O(p^{-3/2})$ and the vector $A_0 = (0, A^T)^T$ is independent of p. Hence, the non-random vector $\boldsymbol{\theta}_0 \to 0$ and the estimator $B + tB_0$ with a fixed t has its age effect estimates $\tilde{\boldsymbol{\theta}}_p + t\boldsymbol{\theta}_0 \to_p \boldsymbol{\theta}^\infty$.* ∎

Remark 8.3

1) Corollary 1 can be achieved with a different penalty $-\lambda[b^T((1/\|B\|)B - (\|B\|/t)B_0)]^2$ in the penalized log-likelihood (8.2) with a fixed $t \neq 0$ independent of p and the intrinsic estimator B for a given data set. From a Bayesian point of view, such a penalty induces a multivariate normal prior distribution on the parameters b, which leads to unique parameter estimation.

2) Theorem 8.1 and Corollary 1 provide an affirmative answer to the identification problem: there exists a unique set of a parameters that the age effects of all multiple estimators converge to asymptotically. By symmetry, it is also true for the period effects. The next theorem provides a summary.

Theorem 8.2 *There exists a unique set of true parameter values of the intercept, age and period effects.*

Now, I consider the $\sqrt{p}$-consistency and asymptotic normality of the estimators B and $B + tB_0$ with a fixed $t \neq 0$.

Theorem 8.3 *Under the regularity conditions, the MaPLE $\tilde{\boldsymbol{\theta}}_p$ by the intrinsic estimator B of the generalized linear model (8.1) converges in probability to* *the intercept and age effects with $\sqrt{p}$-consistency*

$$\sqrt{p}\left(\tilde{\boldsymbol{\theta}}_p - \boldsymbol{\theta}^\infty\right) \to_d N\left(\mathbf{0}, C_1^{-1}\right) \qquad \text{as} \quad p \to \infty.$$

Proof *Take the derivative of the profile log-likelihood $Pl_p^\lambda(\boldsymbol{\theta})$ with respect to $\boldsymbol{\theta}$,*

$$\frac{\partial Pl_p^\lambda(\boldsymbol{\theta})}{\partial \boldsymbol{\theta}} = \frac{\partial l_p^\lambda(\boldsymbol{\theta}, \boldsymbol{\xi})}{\partial \boldsymbol{\theta}} + \frac{\partial l_p^\lambda(\boldsymbol{\theta}, \boldsymbol{\xi})}{\partial \boldsymbol{\xi}} \frac{\partial \boldsymbol{\xi}}{\partial \boldsymbol{\theta}} = \frac{\partial l_p^\lambda(\boldsymbol{\theta}, \boldsymbol{\xi})}{\partial \boldsymbol{\theta}}.$$

Replace $\tilde{\boldsymbol{\theta}}_{p+n}$ with $\boldsymbol{\theta}$ in (8.6) and take partial derivative with respect to $\boldsymbol{\theta}$ at $\boldsymbol{\theta} = \boldsymbol{\theta}^\infty$. For large p,

$$\frac{1}{p} \left.\frac{\partial^2 Pl_p^\lambda(\boldsymbol{\theta}; \boldsymbol{y})}{\partial \boldsymbol{\theta}^2}\right|_{\boldsymbol{\theta}=\tilde{\boldsymbol{\theta}}_p} \sqrt{p}\left(\boldsymbol{\theta}^\infty - \tilde{\boldsymbol{\theta}}_p\right) = \frac{1}{\sqrt{p}} \left.\frac{\partial Pl_p^\lambda(\boldsymbol{\theta}; \boldsymbol{y})}{\partial \boldsymbol{\theta}}\right|_{\boldsymbol{\theta}=\boldsymbol{\theta}^\infty}$$

$$= \frac{1}{\sqrt{p}} \left.\frac{\partial l(\boldsymbol{\theta}, \boldsymbol{\xi}; \boldsymbol{y})}{\partial \boldsymbol{\theta}}\right|_{\boldsymbol{\theta}=\boldsymbol{\theta}^\infty} - \frac{2\lambda}{\sqrt{p}}(\boldsymbol{\theta}^{\infty T}\boldsymbol{\theta}_0 + \boldsymbol{\xi}_{\boldsymbol{\theta}}^T \boldsymbol{\xi}_0)\boldsymbol{\theta}_0. \qquad (8.9)$$

The first term on the right-hand side of (8.9) is the derivative of the log-likelihood function $l(\boldsymbol{\theta}; \boldsymbol{y})$ of a reduced row effect model, whose maximum likelihood estimator is $\sqrt{p}$-consistent and approximately follows a Gaussian distribution $N(\mathbf{0}, C_1^{-1})$ under the regularity conditions. The second term $\Delta = -(2\lambda/\sqrt{p})(\boldsymbol{\theta}^{\infty T}\boldsymbol{\theta}_0 + \boldsymbol{\xi}_{\boldsymbol{\theta}}^T \boldsymbol{\xi}_0)\boldsymbol{\theta}_0$ is a non-random a-dimensional vector. By the regularity conditions, the components of $\boldsymbol{\xi}_{\boldsymbol{\theta}}$ and $\boldsymbol{\xi}_0$ are uniformly bounded. Thus, $(\boldsymbol{\theta}^{\infty T}\boldsymbol{\theta}_0 + \boldsymbol{\xi}_{\boldsymbol{\theta}}^T \boldsymbol{\xi}_0) = O(p)$ and $\Delta = -(2\lambda/\sqrt{p})O(p)\boldsymbol{\theta}_0 = O(1/p)\lambda \boldsymbol{A}_0 \to \mathbf{0}$ since $\boldsymbol{\theta}_0 = O(\sqrt{6}\ p^{-3/2}\boldsymbol{A})$, as shown in the proof of Corollary 1. Alternatively, $\Delta \to \mathbf{0}$ because λ is infinitesimal. By Lemma 3, $\sqrt{p}\,(\tilde{\boldsymbol{\theta}}_p - \boldsymbol{\theta}^\infty) \to_d N\left(\mathbf{0},\ C_1^{-1}\right)$ as $p \to \infty$. ∎

Theorem 8.4 *Under the regularity conditions, the estimator $\boldsymbol{B} + t\boldsymbol{B}_0$ yields $\sqrt{p}$-consistent estimation of the intercept and age effects for a fixed $t \neq 0$ with asymptotically normal distribution*

$$\sqrt{p}\left(\tilde{\boldsymbol{\theta}}_p + t\boldsymbol{\theta}_0 - \boldsymbol{\theta}^\infty\right) \to_d N\left(\mathbf{0}, C_1^{-1}\right) \qquad \text{as} \quad p \to \infty.$$

The proof is straightforward following the $\sqrt{p}$-consistency of $\tilde{\boldsymbol{\theta}}_p$ in Theorem 8.3 and the order of the unit length eigenvector $\boldsymbol{B}_0 = O(p^{-3/2})$, as noticed in the proof of Corollary 1.

Remark 8.4

1) Theorems 8.1–8.4 imply that all multiple estimators with a fixed t converge to the same set of true parameters as p diverges to infinity although the multiple estimators differ largely with finite samples. However, this result is true only for a fixed t. If t varies with p, $\boldsymbol{B} + t\boldsymbol{B}_0$ may yield inconsistent estimation, as shown in the next subsection.

2) With the consistency and asymptotic normality, the estimator $\tilde{\boldsymbol{\theta}}_p$ is expected to yield estimates close to the true parameter values even for moderate sample sizes. This has been illustrated with intensive simulation studies in Fu (2016).

3) The decomposition of estimators (6.1) by eigen analysis implies that $\boldsymbol{B}$ can be computed with the principal component analysis (PCA) method by simply taking all principal components with nonzero eigenvalues and transforming the coefficients of each principal component back to age, period and cohort effects; see Fu (2008) for details. In fact, the intrinsic estimator is computed through the PCA approach in the **R** program.

4) Although the intrinsic estimator is computed through the PCA approach, the variance estimation provided with the PCA approach may not be appropriate, not only because the reduced model by the MaPLE estimator has different degrees of freedom, but also because the numbers of the periods and cohorts diverge to infinity, leaving the variance estimation of the period and cohort effects by the PCA approach invalid for non-Gaussian response variables. I will address the variance estimation issue in Chapter 9.

5) For the case with a diverging number of age groups $a = a(p)$ for large p, the above asymptotic results need to be updated to allow a diverging number of age groups in addition to the diverging number of periods. Interested readers are referred to the results in Fu and Hall (2006) for the asymptotics with Gaussian responses and diverging numbers of age groups and periods.

8.3.2 Asymptotics of Linearly Constrained Estimators

I now consider the asymptotics of the multiple estimators with varying t. In particular, I study constrained estimators, focusing on linear constraints. Note that a constraint is often in a linear form, such as $\boldsymbol{l}^T \boldsymbol{b} = 0$, of which the equality constraints are special cases. Theorem 8.4 offers consistency of an estimator $\boldsymbol{B} + t\boldsymbol{B}_0$ with a fixed $t \neq 0$. But a constraint may induce a non-zero t that varies with p. Hence the large sample behavior of the estimators depend not only on the consistency of the intrinsic estimator $\boldsymbol{B}$, corresponding to $t = 0$, but also on the behavior of the null vector component $t\boldsymbol{B}_0$ with a varying t, whose behavior differs largely from the one with a fixed t. Since a constraint $\boldsymbol{l}^T \boldsymbol{b} = 0$ can be set on age, period, or cohort effects, but not on the intercept, I abuse the notation and denote by $\boldsymbol{l}^T \boldsymbol{\alpha}$, $\boldsymbol{l}^T \boldsymbol{\beta}$, and $\boldsymbol{l}^T \boldsymbol{\gamma}$ a constraint solely on the age groups, periods and cohorts, respectively. In addition, I need to exclude a special class of constraints since they do not achieve full rank on the singular matrix X.

Lemma 4 *If a linear constraint $\boldsymbol{l}^T \boldsymbol{b} = 0$ on matrix X satisfies simultaneously $\boldsymbol{l}^T \boldsymbol{A} = 0$, $\boldsymbol{l}^T \boldsymbol{P} = 0$ and $\boldsymbol{l}^T \boldsymbol{C} = 0$, the constraint model has the same null space spanned by the null vector $\{t\boldsymbol{v}\}$ of matrix X.*

Proof *Notice the special closed form of the eigenvector $\boldsymbol{v} = (0, \boldsymbol{A}^T, \boldsymbol{P}^T, \boldsymbol{C}^T)^T$. A linear constraint satisfies $\boldsymbol{l}^T \boldsymbol{v} = 0$ if it satisfies all equations $\boldsymbol{l}^T \boldsymbol{A} = 0$, $\boldsymbol{l}^T \boldsymbol{P} = 0$ and $\boldsymbol{l}^T \boldsymbol{C} = 0$. Hence vector $\boldsymbol{l}$ and all column vectors of the design* *matrix X are perpendicular to $\boldsymbol{v}$. $\boldsymbol{l}$ is thus a linear combination of the column vectors of matrix X because $\boldsymbol{v}$ is the only eigenvector of $X^T X$ with an eigen value of 0. Hence, the constraint $\boldsymbol{l}^T \boldsymbol{b} = 0$ does not change the null space of X.*

∎

8.3.2.1   Linear Constraint on Age Effects

Consider a constraint $\boldsymbol{l}^T \boldsymbol{\alpha} = 0$ on the age effects, e.g. $\alpha_1 = \alpha_3$, based on prior knowledge of the event under investigation. One expects the identity to be true for the current $a \times p$ table as well as a larger $a \times q$ table ($q > p$) with more data in the extra $(q-p)$ columns. The constraint implies $\boldsymbol{l}^T (\tilde{\boldsymbol{\theta}}_p + t \boldsymbol{A} / \|\boldsymbol{v}\|) = 0$ and hence $t = -\boldsymbol{l}^T \tilde{\boldsymbol{\theta}}_p \|\boldsymbol{v}\| / (\boldsymbol{l}^T \boldsymbol{A})$.

Note that the vectors $\boldsymbol{A}_0 = (0\ \boldsymbol{A}^T)^T$ and $\boldsymbol{\theta}_0 = \boldsymbol{A}_0 / \|\boldsymbol{v}\|$ with the latter being a sub-vector of $\boldsymbol{B}_0$. For a given constraint $\boldsymbol{l}^T \boldsymbol{\alpha} = 0$, the intercept and age effect estimates follow

$$\sqrt{p}\left((\tilde{\boldsymbol{\theta}}_p + t\boldsymbol{\theta}_0) - (\boldsymbol{\theta}^\infty - \frac{\boldsymbol{l}^T \boldsymbol{\theta}^\infty}{\boldsymbol{l}^T \boldsymbol{A}} \boldsymbol{A}_0) \right) = \sqrt{p}\,(\tilde{\boldsymbol{\theta}}_p - \boldsymbol{\theta}^\infty) - \frac{\boldsymbol{l}^T \sqrt{p}\,(\tilde{\boldsymbol{\theta}}_p - \boldsymbol{\theta}^\infty)}{\boldsymbol{l}^T \boldsymbol{A}} \boldsymbol{A}_0 .$$

$$(8.10)$$

By $\sqrt{p}$-consistency of the MaPLE $\tilde{\boldsymbol{\theta}}_p$, both terms on the right-hand side of (8.10) follow an asymptotically normal distribution with mean $\boldsymbol{0}$. Hence, the estimates converge to a limiting vector $\boldsymbol{\theta}^\infty - \boldsymbol{A}_0 \boldsymbol{l}^T \boldsymbol{\theta}^\infty / (\boldsymbol{l}^T \boldsymbol{A})$. Note that vectors $\boldsymbol{A}$ and $\boldsymbol{A}_0$ do not depend on p. Denote by $\boldsymbol{\theta}_A^\infty$ the age effect vector in $\boldsymbol{\theta}^\infty$.

Theorem 8.5 *Under the regularity conditions, a linear constraint on age effects $\boldsymbol{l}^T \boldsymbol{\alpha} = 0$ yields age effect estimates $\tilde{\boldsymbol{\alpha}}_p$ with*

$$\sqrt{p}\left(\tilde{\boldsymbol{\alpha}}_p - (\boldsymbol{\theta}_A^\infty - \frac{\boldsymbol{l}^T \boldsymbol{\theta}_A^\infty}{\boldsymbol{l}^T \boldsymbol{A}} \boldsymbol{A}) \right) \longrightarrow_d N(\boldsymbol{0}, \Sigma_A) \ \ as \ \ p \to \infty$$

for some positive definite $(a - 1) \times (a - 1)$ variance-covariance matrix Σ_A. Hence,

1) if $\boldsymbol{l}^T \boldsymbol{\theta}_A^\infty = 0$, i.e., the constraint is satisfied by the limiting age effects $\boldsymbol{\theta}_A^\infty$,

$\tilde{\boldsymbol{\alpha}}_p$ *converges to the limit* $\boldsymbol{\theta}_A^\infty$ *with asymptotic normality;*

2) if $\boldsymbol{l}^T \boldsymbol{\theta}_A^\infty \neq 0$, *i.e., the constraint is not satisfied by the limiting age effects, the constraint yields age effect estimates with asymptotic bias* $-\boldsymbol{l}^T \boldsymbol{\theta}_A^\infty \boldsymbol{A}/(\boldsymbol{l}^T \boldsymbol{A})$.

Proof *Following the* $\sqrt{p}$-*consistency of the MaPLE* $\tilde{\boldsymbol{\theta}}_p$ *in Theorem 8.3, both terms on the right-hand side of equation (8.10) follow an approximately normal distribution with mean 0.* ∎

Remark 8.5 Theorem 8.5 makes the role of the limit $\boldsymbol{\theta}_A^\infty$ clear. If a constraint is satisfied by the limit, it yields consistent estimation of the age effects. In contrast, if a constraint is not satisfied by the limit, it yields age effect estimates with asymptotic bias. By the rationale of the constraint approach, the limiting vector $\boldsymbol{\theta}_A^\infty$ functions as the true age effect parameters. Hence, it justifies the existence and uniqueness of the true age effect parameters. By Theorem 8.2, it further implies that there exists a set of true parameters of the age and period effects. Consequently, a set of true parameters of the cohort effects also exists, assuming the APC models are correct.

8.3.2.2 Linear Constraint on Period or Cohort Effects

By symmetry, similar results on the consistency and asymptotic normality of the period effect estimates are also true as the number of age groups a diverges to infinity. However, more interesting results can be derived with linear constraints on the period or cohort effects as p diverges to infinity. To provide further results on the asymptotics, I introduce a concept of contrast constraint.

Definition *A linear constraint* $\boldsymbol{l}^T \boldsymbol{b} = 0$ *on the parameters* $\boldsymbol{b}$ *is a contrast constraint if the sum* $\sum l_j = 0$.

I examine the behavior of age effect estimates by a linear constraint on the period effects $l^T\beta = 0$ or on the cohort effects $l^T\gamma = 0$ as $p \to \infty$. Assume the constraint has nonzero elements $l_j \neq 0$ only in the first L elements (L is bounded; $L < p$ for periods or $L < a + p - 1$ for cohorts) so that the same parameters β_j or γ_k remain in the constraint as p increases. For example, $l^T b = 0$ does not vary with p as in $\beta_1 = \beta_2$, but varies in $\beta_1 = \beta_p$. The constraint implies $t = -l^T \tilde{\xi}_P^\lambda \|v\| / l^T P$, or $t = -l^T \tilde{\xi}_C^\lambda \|v\| / l^T C$, where $\tilde{\xi}_P^\lambda$ and $\tilde{\xi}_C^\lambda$ are profile vectors for the period and cohort effect estimates, respectively, by B.

The constraint $l^T\beta = 0$ yields the age effect estimates $\tilde{\theta}_p + t\theta_0 = \tilde{\theta}_p - l^T \xi_P^\lambda(\tilde{\theta}) A_0 / (l^T P)$. By the mean value theorem on the profile function $\xi_P^\lambda(\theta)$ with a bounded derivative $\xi_P^{\lambda'}(\theta)$,

$$\sqrt{p}\left[(\tilde{\theta}_p + t\theta_0) - (\theta^\infty - A_0 \frac{l^T \xi_P^\lambda(\theta^\infty)}{l^T P})\right]$$

$$= \sqrt{p}(\tilde{\theta}_p - \theta^\infty) - A_0 \frac{l^T \xi_P^{\lambda'}(\theta^*)\sqrt{p}(\tilde{\theta}_p - \theta^\infty)}{l^T P}. \tag{8.11}$$

By $\sqrt{p}$-consistency of the MaPLE $\tilde{\theta}_p$, both terms on the right-hand side of equation (8.11) follow an asymptotically normal distribution with mean 0. Similar results are true for the age estimates by a constraint on cohort effects. Note that both vectors P and C depend on p in a special form as shown in (6.2).

Theorem 8.6 *Under the regularity conditions, age effect estimates $\tilde{\alpha}_p$ by a constraint on the period or cohort effects follow an approximately normal distribution as $p \to \infty$.*

$$\sqrt{p}\left(\tilde{\alpha}_p - \left[\theta_A^\infty - A\frac{l^T \xi_P^\lambda(\theta^\infty)}{l^T P}\right]\right) \longrightarrow_d N(0, \Sigma_P)$$

with a constraint on periods $l^T\beta = 0$, or

$$\sqrt{p}\left(\tilde{\alpha}_p - \left[\theta_A^\infty - A\frac{l^T \xi_C^\lambda(\theta^\infty)}{l^T C}\right]\right) \longrightarrow_d N(0, \Sigma_C)$$

with a constraint on cohorts $l^T \gamma = 0$, for some positive-definite variance-covariance matrix Σ_P or Σ_C of dimension $(a-1) \times (a-1)$. Furthermore,

1) if $l^T \xi^\lambda(\theta^\infty) = 0$, i.e. the constraint is satisfied by true period or cohort effects as a profile function of the true age effects, $\tilde{\alpha}_p$ is then $\sqrt{p}$-consistent.

2) if $l^T \xi^\lambda(\theta^\infty) \neq 0$, i.e. the constraint is not satisfied by true period or cohort effects as a profile function of the true age effects, a contrast constraint yields asymptotic bias $-A l^T \xi^\lambda(\theta^\infty)/l^T \xi_0$ for the age effects, where $l^T \xi_0 = l^T P$ for the constraint on periods or $l^T \xi_0 = l^T C$ for the constraint on cohorts, and neither depends on p.

3) If $l^T \xi^\lambda(\theta^\infty) \neq 0$, a non-contrast constraint yields consistent estimates because the bias

$$-A l^T \xi^\lambda(\theta^\infty)/l^T \xi_0 = O(p^{-1}) \to 0 \text{ as } p \to \infty.$$

Proof   *Following the $\sqrt{p}$-consistency of the MaPLE $\tilde{\theta}_p$ in Theorem 8.3, both terms on the right-hand side of equation (8.11) follow an approximately normal distribution with mean 0. Notice the special form of vectors P and C. If l is a contrast, $l^T P = (-1, \ldots, -(p-1)) l$ does not depend on p since $l_j = 0$ for $j \geq L$ with bounded $L < p-1$, i.e. l has nonzero entries only in the first finite number of components. Similarly, $l^T C$ does not depend on p for a contrast l. If l is not a contrast, $l^T P = O((\sum l_j)(p-1)/2)) = O(p)$ and $l^T C = O(p)$, which implies the consistency.*   ∎

Remark 8.6

1) Since the limiting vector θ^∞ functions as the true age effect parameter vector, $\xi^\lambda(\theta^\infty)$ represents period and cohort effects by the profile function of true age effect parameters. Following a similar discussion to that of true age effect parameters, the profiled period and cohort effects $\xi^\lambda(\theta^\infty)$ play the same role as true period and cohort effects by the rationale of the constraint

approach. That is, a constraint yields consistent estimation of the age effects if it is satisfied by the profiled period or cohort effects as a profile function of the true age effects, but yields asymptotically biased estimates of the age effects otherwise. Hence, the profiled period and cohort effects function as the unique parameters of true period and cohort effects.

2) In practice, it is difficult to assess the validity of a constraint satisfied by true parameters and virtually impossible to search for a constraint, because the true parameters are unknown. However, Theorem 8.6 makes the consistency easy to achieve by setting a non-contrast constraint on period or cohort effects, such as $\gamma_1 = 2\gamma_4$. Such non-contrast constraints have never been used in practice since they are not as intuitive as equality constraints.

3) Since the popular equality constraints are contrasts, they do not yield consistent estimation by Theorem 8.6 unless they are satisfied by true parameters of age, period, or cohort effects, which is almost surely impossible. This explains why an equality constraint often yields insensible trend estimation, no matter how reasonable the assumption for the equality constraint is. The equality constraints do not help to resolve the identification problem as long as the true parameters remain unknown, even if the investigator has full knowledge of the event under investigation.

4) Although non-contrast constraints on period or cohort effects yield consistent estimation of age effects with the bias converging to 0, the convergence speed is moderate at the order $O(1/p)$, which is numerically observable by comparing non-contrast constraint and the intrinsic estimator.

5) Theorems 8.5 and 8.6 provide asymptotic bias of inconsistent estimators though accurate estimation of the bias depends on true parameter values, which may often be approximated by consistent estimates, such as those by the intrinsic estimator.

6) Based on the above consistency results on the estimators $B + tB_0$ with either a fixed t or a varying t by a linear constraint, one knows which estimators yield consistent estimation and which ones do not. Hence, even without *a priori* knowledge of the event under investigation, one can choose consistent estimators for the best results in data analysis. In such a case, the intrinsic estimator yields consistent estimation of the age effects, and performs the best among the infinitely many estimators.

8.4   Estimability of Intrinsic Estimator

In this section, I provide a rigorous proof of the estimability of the intrinsic estimator B. Notice that following linear model theory (Scheffé 1959), a linear combination $l^T b$ of model parameters b is estimable if it is the expected value of a linear combination of the responses y. A sufficient condition is that the coefficient vector of the parameters in the estimable function can be written as a linear combination of the row vectors of the design matrix X. This can be easily extended to multiple estimable functions, where the product of a matrix by the model parameters is considered. I thus give the following theorem.

Theorem 8.7   *The expected value of the intrinsic estimator B is an estimable function of the APC model (4.4).*

Proof   *Following a sufficient condition for estimable function (Scheffé 1959), it suffices to prove that there exists a matrix U such that $(I - B_0 B_0^T) = UX$. It further suffices to prove that there exists a matrix U_0 such that $U_0 X^T X = (I - B_0 B_0^T)$ since matrix $U = U_0 X^T$ will serve the purpose.*

Consider a linear transformation $\mathcal{X} : \mathbb{R}^m \to \mathbb{R}^m$ induced by the matrix $X^T X$. To prove the existence of such a matrix U_0, it suffices to prove that the matrix $(I - \boldsymbol{B_0}\boldsymbol{B_0}^T) \in Image(\boldsymbol{\mathcal{X}})$. By the Rank-Nullity Theorem (Bretscher 2004) $dim(\mathbb{R}^m) = dim(Kernel(\boldsymbol{\mathcal{X}})) + dim(Image(\boldsymbol{\mathcal{X}}))$ and $\mathbb{R}^m = Kernel(\boldsymbol{\mathcal{X}}) \oplus Image(\boldsymbol{\mathcal{X}})$, it suffices to prove that $(I - \boldsymbol{B_0}\boldsymbol{B_0}^T) \notin Kernel(\boldsymbol{\mathcal{X}})$. This is verified by $X^TX(I - \boldsymbol{B_0}\boldsymbol{B_0}^T) = X^TX - X^TX\boldsymbol{B_0}\boldsymbol{B_0}^T = X^TX \neq 0$, i.e. no column vector is $\boldsymbol{0}$. Thus $(I - \boldsymbol{B_0}\boldsymbol{B_0}^T) \in Image(\boldsymbol{\mathcal{X}})$, i.e. all column vectors, and all row vectors by symmetry, are in the image of the transformation $\boldsymbol{\mathcal{X}}$. Thus there exists a matrix U_0 such that $U_0 X^TX = (I - \boldsymbol{B_0}\boldsymbol{B_0}^T)$. Hence $\boldsymbol{b}_1 = (I - \boldsymbol{B_0}\boldsymbol{B_0}^T)\boldsymbol{b}$ is estimable.

To prove the uniqueness, notice that an estimator that completely determines the parameters must be of the form in (6.1). Hence an estimable function that completely determines the parameters is of the form $LE(\hat{\boldsymbol{b}}) = L(E(\boldsymbol{B}) + t\boldsymbol{B_0}) = L\boldsymbol{b}_1 + tL\boldsymbol{B_0}$. By the estimability, $LE(\hat{\boldsymbol{b}})$ is invariant and independent of the arbitrary number t (Searle 1971). Thus, $L\boldsymbol{B_0} = \boldsymbol{0}$, which implies $LE(\hat{\boldsymbol{b}}) = L\boldsymbol{b}_1$, i.e. such estimable functions must have no component of $\boldsymbol{B_0}$. Consequently, $E(\boldsymbol{B} + s\boldsymbol{B_0}) = \boldsymbol{b}_1 + s\boldsymbol{B_0}$ with $s \neq 0$ is not estimable, leaving $\boldsymbol{b}_1$ the only estimable function.   ∎

Remark   The above proof of the estimability confirms that $E\boldsymbol{B}$ is estimable. Consequently, it has an unbiased estimator $\boldsymbol{B}$. Following the estimability concept, $\boldsymbol{B}$ does not depend on the constraint that is used to yield a unique estimator. Notice that this estimability is within the framework of APC models in which the side condition is already selected, such as the centralization on age, period, and cohort effects. It has been reported in a recent work that as the side condition changes, such as from the centralization to a reference level $\alpha_1 = \beta_1 = \gamma_1 = 0$, the intrinsic estimator $\boldsymbol{B}$ varies (Luo et al 2016). This is an interesting observation, which implies that the selection of side condition is crucially important in APC modeling and deserves more attention than before. However, it does not disqualify $E\boldsymbol{B}$ being estimable because the identification problem discussed in the literature is on the parameter identification within the setting of a selected side condition, such as the centralization. The selection of the side condition is an important yet different issue, and will be discussed in Chapter 9, where I derive theoretical results and conduct simulations to conclude that the centralization is the best side condition recommended for APC models.

8.5 Suggested Readings

The asymptotics of multiple estimators have rarely been studied before due to the difficulty in working with multiple estimators. There are only a few papers on this topic. The penalty models provide an efficient approach to achieving a unique estimator, see Fu (2000) for the application of ridge penalty. Knight and Fu (2000) also studied the asymptotics of Lasso-type estimators with linearly dependent covariates. Fu (2008) studied the asymptotics by applying a smoothing technique to the cohort effects to achieve a unique estimator. Fu (2016) studied the asymptotics of the APC multiple estimators with a fixed number of age groups. Fu and Hall (2006) studied the asymptotics with diverging number of age groups and periods for a linear model.

The following techniques are important in the study of asymptotics in the APC models. The work by Neyman and Scott (1948) on the consistent estimation of the model variance component of an $a \times p$ table is a good motivation, and demonstrates that the uniqueness of an estimator is not crucial, but the consistency of the estimator is. The monograph by McCullagh and Nelder (1989) is one of the best reference books for the generalized linear models.

Murphy and van der Vaart (2000) studied the profile likelihood and provides a general guide to the work on profile likelihood.

8.6   Exercises

1.* Why is the assumption of the infinitesimal tuning parameter λ valid for the definition of the penalized profile log-likelihood?

2.* Why is the assumption of the infinitesimal parameter λ important for the asymptotic studies?

3.* Without taking the profile likelihood approach, examine the asymptotic behavior of the maximum likelihood estimator of the penalized log-likelihood function. Can you achieve similar asymptotic results on the parameters?

4.* Profile likelihood is not likelihood and may lead to biased estimation. Why does the employment of profile likelihood in the APC models yield consistent estimation?

---

* Difficult exercises with an asterisk are meant for graduate students in biostatistics.

9

Variance Estimation and Selection of Side Conditions

In Chapters 6 and 8, I provided the rationale and theoretical justification with full details on why the intrinsic estimator presents reasonable results, and why the popular equality constraints yield biased estimation even though the specification of the constraints often carries reasonable assumptions. In this chapter, I study two important issues of parameter estimation. One is to accurately calculate variances and standard errors based on the asymptotic results and establish the Delta method for the variance of the period and cohort effects in fitting a generalized linear model to the APC data. The other is to select the side conditions by searching for efficient estimation with least variance through theoretical calculation and simulation studies.

9.1 Variance Estimation of the Intrinsic Estimator

Although the parameter estimates of the intrinsic estimator can be efficiently computed using the PCA method for both the linear and loglinear APC models, the variance estimation from the PCA approach may not be valid for the loglinear models. By the large sample theory, the age effect estimates of the intrinsic estimator follow an approximately normal distribution as the number of periods p diverges to infinity. Hence, assuming the responses in the loglinear models follow a Poisson distribution, the asymptotic distribution of the age effects follows a normal distribution derived from the maximum profile likelihood estimates of the age effects. However, this may not be true for the period and cohort effect estimates as the number of parameters for the period and cohort effects diverge to infinity. This implies that the derivation of the asymptotic variance of the period and cohort effects by the PCA approach may be invalid.

9.1.1 The Delta Method for the Variance of Period and Cohort Effect Estimates

Following the asymptotic results in Chapter 8, the intercept and age effects estimates $\widehat{\boldsymbol{\theta}}$ by the intrinsic estimator have an approximately normal distribution with

$$\widehat{\boldsymbol{\theta}} \sim N \left(\boldsymbol{\theta}^{\infty}, \ \frac{C_1^{-1}}{p} \right) \quad \text{as } p \longrightarrow \infty .$$

Since the period and cohort effect estimates are profile functions of the age effects by the maximum profile likelihood method, as shown in the asymptotic studies in Chapter 8, I use the Delta method below to derive the variance of the period and cohort effect estimates $\widehat{\boldsymbol{\xi}}$.

First, I introduce the Delta method, which is widely used in statistics. Suppose a random variable Y has finite mean $E(Y)$ and finite variance $Var(Y)$. Assume f is a real differentiable function of the random variable Y. The Taylor expansion yields

$$f(Y) = f(EY) + f'(EY)(Y - EY) + o(Y - EY)$$

Taking the variance on both sides of the above equation provides an approximation to the variance of $f(Y)$;

$$Var(f(Y)) \approx [f'(EY)]^2 Var(Y) .$$

For notational convenience, I omit the hat symbol ($\hat{\ }$) in the effect estimates. Let $\boldsymbol{\xi} = \boldsymbol{\xi_\theta}$ be a nonlinear function of $\boldsymbol{\theta}$, and denote by X_1 and X_2 the design matrices corresponding to the age effects $\boldsymbol{\theta}$ and the period and cohort effects $\boldsymbol{\xi}$, respectively. By the Delta method,

$$Var(\boldsymbol{\xi}) = Var(\boldsymbol{\xi_\theta}) = (\frac{\partial \boldsymbol{\xi}}{\partial \boldsymbol{\theta}})Var(\boldsymbol{\theta})(\frac{\partial \boldsymbol{\xi}}{\partial \boldsymbol{\theta}})^T \tag{9.1}$$

By Remark 8.1, the penalty parameter $\lambda > 0$ is infinitesimal. Hence, in the derivation of the Delta method, the penalty term is also infinitesimal and does not contribute to the variance estimation. For this reason, the penalized profile likelihood is equivalent to the profile likelihood.

$$\frac{\partial l^\lambda(\boldsymbol{\theta}, \boldsymbol{\xi}; \boldsymbol{y})}{\partial \boldsymbol{\xi}} = 0, \forall \boldsymbol{\theta}.$$

Taking the derivative with respect to $\boldsymbol{\theta}$ on both sides implies

$$\frac{\partial l^2(\boldsymbol{\theta}, \boldsymbol{\xi}; \boldsymbol{y})}{\partial \boldsymbol{\xi} \partial \boldsymbol{\theta}} + \frac{\partial l^2(\boldsymbol{\theta}, \boldsymbol{\xi}; \boldsymbol{y})}{\partial \boldsymbol{\xi}^2}\frac{\partial \boldsymbol{\xi}}{\partial \boldsymbol{\theta}} = 0,$$

which leads to

$$\frac{\partial \boldsymbol{\xi}}{\partial \boldsymbol{\theta}} = -\left[\frac{\partial l^2(\boldsymbol{\theta}, \boldsymbol{\xi}; \boldsymbol{y})}{\partial \boldsymbol{\xi}^2}\right]^{-1}\frac{\partial l^2(\boldsymbol{\theta}, \boldsymbol{\xi}; \boldsymbol{y})}{\partial \boldsymbol{\xi} \partial \boldsymbol{\theta}}.$$

I demonstrate how to apply the Delta method to calculate the variance for the profile parameter $\boldsymbol{\xi}$ with two distributions in the exponential family: the Gaussian distribution and Poisson distribution.

Denote by $\boldsymbol{b}^T = (\boldsymbol{\theta}^T, \boldsymbol{\xi}^T)$ the parameters of the exponential family distribution for the response Y. The log-likelihood function is $l(\boldsymbol{\theta}, \boldsymbol{\xi}; \boldsymbol{y}) = [\boldsymbol{y}^T \boldsymbol{\zeta}(\boldsymbol{\theta}, \boldsymbol{\xi}) - \psi(\boldsymbol{\zeta}(\boldsymbol{\theta}, \boldsymbol{\xi}))]/a(\Phi) + c(\boldsymbol{y}, \Phi)$. Furthermore, the penalized log-likelihood $l^\lambda(\boldsymbol{\theta}, \boldsymbol{\xi}; \boldsymbol{y}) = l(\boldsymbol{\theta}, \boldsymbol{\xi}; \boldsymbol{y}) - \lambda(\boldsymbol{b}^T \boldsymbol{B}_0)^2 = l(\boldsymbol{\theta}, \boldsymbol{\xi}; \boldsymbol{y}) + O(\lambda)$ with infinitesimal $\lambda > 0$. Thus, the last term is omitted in the following calculation for the Delta method.

Case 1. Gaussian distribution $l(\boldsymbol{\zeta}; \boldsymbol{y}) = [\boldsymbol{y}^T \boldsymbol{\zeta} - \psi(\boldsymbol{\zeta})]/a(\Phi) + c(\boldsymbol{y}, \Phi)$. For the canonical link, $\zeta = g(\boldsymbol{\mu}) = \boldsymbol{\eta} = X\boldsymbol{b}$ with $g(\boldsymbol{\mu}) = \boldsymbol{\mu}$. Hence,

$$\frac{\partial l}{\partial \zeta} = [\boldsymbol{y}^T - \boldsymbol{\mu}]/a(\Phi) = [\boldsymbol{y} - X\boldsymbol{b}]/a(\Phi)\,, \quad \frac{\partial l}{\partial \boldsymbol{b}} = \frac{\partial l}{\partial \zeta}\frac{\partial \zeta}{\partial \boldsymbol{b}} = X^T[\boldsymbol{y} - X\boldsymbol{b}]/a(\Phi)\,,$$

and

$$\frac{\partial^2 l}{\partial \boldsymbol{\xi}^2} = (X_2^T X_2)/a(\Phi)\,, \quad \frac{\partial^2 l}{\partial \boldsymbol{\xi}\partial \boldsymbol{\theta}} = (X_2^T X_1)/a(\Phi)\,,$$

which leads to

$$\left(\frac{\partial^2 l}{\partial \boldsymbol{\xi}^2}\right)^{-1}\frac{\partial^2 l}{\partial \boldsymbol{\xi}\partial \boldsymbol{\theta}} = (X_2^T X_2)^{-1}(X_2^T X_1)\,.$$

It is thus implied that

$$Var(\boldsymbol{\xi_\theta}) = (X_2^T X_2)^{-1}(X_2^T X_1)Var(\boldsymbol{\theta})(X_1^T X_2)(X_2^T X_2)^{-1}\,.$$

For a given asymptotic variance $Var(\boldsymbol{\theta}) = C_1^{-1}/p$ for large p, by Theorem 8.3, the variance of the profile estimator by the Delta method is given by

$$Var(\boldsymbol{\xi_\theta}) = \frac{1}{p}(X_2^T X_2)^{-1}(X_2^T X_1)C_1^{-1}(X_1^T X_2)(X_2^T X_2)^{-1}\,.$$

Since the rows and columns of a given APC data set are orthogonal and their effects are independent of each other, the column vectors of the age effects in the design matrix are orthogonal to those of the period effects. Let X_{10} and X_{20} be their corresponding matrices in the design matrix, then $X_{10}^T X_{20} = 0$, a zero matrix. Hence, for the linear AP model, the variance of the period effect estimates can be shown to be independent of the age effects through straightforward calculation based on the above $Var(\boldsymbol{\xi_\theta})$. It is thus recommended that the variances of the period and cohort effects be calculated based on the normal distribution assumption of the linear model, requiring no profile likelihood.

Case 2. Poisson distribution For the canonical link, $\zeta = g(\boldsymbol{\mu}) = \boldsymbol{\eta} = X\boldsymbol{b}$ with $g(\boldsymbol{\mu}) = log(\boldsymbol{\mu})$ and $\boldsymbol{\mu} = g^{-1}(\boldsymbol{\eta}) = \exp(\boldsymbol{\eta})$. Hence,

$$\frac{\partial l}{\partial \zeta} = [\boldsymbol{y}^T - \boldsymbol{\mu}]/a(\Phi) = [\boldsymbol{y} - g^{-1}(X\boldsymbol{b})]/a(\Phi)\,,$$

$$\frac{\partial l}{\partial \boldsymbol{b}} = \frac{\partial l}{\partial \zeta}\frac{\partial \zeta}{\partial \boldsymbol{b}} = X^T[\boldsymbol{y} - g^{-1}(X\boldsymbol{b})]/a(\Phi)$$

and

$$\frac{\partial^2 l}{\partial \boldsymbol{\xi}^2} = (X_2^T \exp(X\boldsymbol{b})X_2)/a(\Phi), \quad \frac{\partial^2 l}{\partial \boldsymbol{\xi} \partial \boldsymbol{\theta}} = (X_2^T \exp(X\boldsymbol{b})X_1)/a(\Phi),$$

where $\exp(X\boldsymbol{b})$ is a diagonal matrix with diagonal elements $\exp(X\boldsymbol{b})$. It then leads to

$$\left(\frac{\partial^2 l}{\partial \boldsymbol{\xi}^2}\right)^{-1}\frac{\partial^2 l}{\partial \boldsymbol{\xi} \partial \boldsymbol{\theta}} = (X_2^T \exp(X\boldsymbol{b})X_2)^{-1}(X_2^T \exp(X\boldsymbol{b})X_1).$$

It is thus implied that

$$Var(\boldsymbol{\xi_\theta}) = (X_2^T \exp(X\boldsymbol{b})X_2)^{-1}(X_2^T \exp(X\boldsymbol{b})X_1)$$
$$\times Var(\boldsymbol{\theta})(X_1^T \exp(X\boldsymbol{b})X_2)(X_2^T \exp(X\boldsymbol{b})X_2)^{-1}$$

For a given asymptotic variance $Var(\boldsymbol{\theta}) = C_1^{-1}/p$ for large p, by Theorem 8.3, the variance of the profile estimator by the Delta method is given by

$$Var(\boldsymbol{\xi_\theta}) = \frac{1}{p}(X_2^T \exp(X\boldsymbol{b})X_2)^{-1}(X_2^T \exp(X\boldsymbol{b})X_1)$$
$$\times C_1^{-1}(X_1^T \exp(X\boldsymbol{b})X_2)(X_2^T \exp(X\boldsymbol{b})X_2)^{-1}.$$

Note that for non-Gaussian distributions, in particular, the Poisson distribution in loglinear models, the matrix $\exp(X\boldsymbol{b})$ is not an identity matrix and thus the orthogonal matrices for the age and period effects do not make the period effect variance $Var(\boldsymbol{\xi_\theta})$ independent of the age effects $\boldsymbol{\theta}$. This helps calculating the variance of the estimates of the period or cohort effects using the asymptotic normality property of the parameter estimate of age effects $\boldsymbol{\theta}$.

Also notice that although the Delta method has been demonstrated with two distributions: the Gaussian and Poisson distributions, the calculation for the Gaussian distribution only serves as an illustration. In practice, it is unnecessary to calculate the standard errors using the Delta method when fitting a linear model to a Gaussian response because the effect estimates follow a normal distribution.

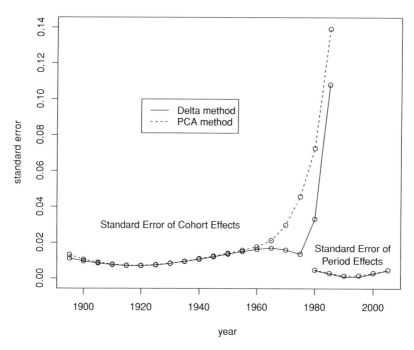

FIGURE 9.1

Comparison of standard errors by the PCA and Delta methods for the period and cohort effects of the intrinsic estimator fitted by the loglinear model to the lung cancer mortality data among US males.

9.1.2 Comparison of Standard Errors between the PCA and Delta Methods

I now compare the standard errors of the intrinsic estimator calculated by the PCA and Delta methods, using the loglinear model fitted to the lung cancer mortality data among US males. Interested readers may find intensive simulation studies in comparing the two methods in Fu et al (2017).

Table 9.1 displays the intrinsic estimator and the standard errors calculated by the PCA and Delta methods for the loglinear model fitted to the lung cancer mortality data among US males. Since the Delta method computes the standard errors of the period and cohort effects based on the age effects, the age effect standard errors remain the same between the two methods. Figure 9.1 presents the standard errors of the period and cohort effects of the intrinsic estimator by the PCA and Delta methods. It is shown that the standard errors by the Delta method are smaller than but close to those by the PCA method, except for a few large values for the young cohort effects, where the Delta method yields much smaller standard errors than the PCA method. Since the over-dispersion may greatly alter the standard errors, I also examine the standard errors by the two methods calculated with the quasi-likelihood approach. Table 9.2 displays the intrinsic estimator and the standard errors by the PCA and Delta methods. Figure 9.2 presents the intrinsic estimator with the confidence intervals by the two methods for the loglinear model with the maximum likelihood approach. In contrast, Figure 9.3 shows the intrinsic estimator with the confidence intervals by the two methods for the loglinear model with the quasi-likelihood approach. In both model fitting approaches, the Delta method produces smaller standard errors than the PCA method. Similar findings can be found in the simulation studies in the aforementioned paper (Fu et al 2017).

TABLE 9.1
Comparison of Standard Errors of the Intrinsic Estimator Between the PCA and Delta Methods with the Loglinear Model Fitted by the Maximum Likelihood Approach to the Lung Cancer Mortality Data among US Males

	Estimate	StdErr.PCA	StdErr.Delta
Intercept	-8.11098305	0.008924914	0.008924914
Age 20	-4.35991764	0.055542226	0.055542226
Age 25	-3.68300089	0.034298516	0.034298516
Age 30	-2.48108478	0.020353635	0.020353635
Age 35	-1.41191334	0.014405935	0.014405935
Age 40	-0.49049301	0.011283546	0.011283546
Age 45	0.21951963	0.009217598	0.009217598
Age 50	0.77307172	0.007526818	0.007526818
Age 55	1.18864017	0.006073351	0.006073351
Age 60	1.49133295	0.004956222	0.004956222
Age 65	1.69547097	0.004408422	0.004408422
Age 70	1.81122490	0.004626446	0.004626446
Age 75	1.83502319	0.005523316	0.005523316
Age 80	1.79487989	0.006860462	0.006860462
Age 85	1.61724625	0.008465306	0.008465306
Period 1980	-0.27591158	0.004802385	0.004588451
Period 1985	-0.13278598	0.003073155	0.002734010
Period 1990	-0.01289789	0.001635008	0.000893252
Period 1995	0.05637571	0.001663657	0.000948635
Period 2000	0.14006679	0.003073750	0.002753575
Period 2005	0.22515294	0.004723509	0.004515796

	Estimate	StdErr.PCA	StdErr.Delta
Cohort 1895	1.25046113	0.013151982	0.011138561
Cohort 1900	1.27564055	0.010481745	0.009662846
Cohort 1905	1.26564590	0.008896570	0.008483065
Cohort 1910	1.20092553	0.007860463	0.007601776
Cohort 1915	1.09093409	0.007301955	0.007107091
Cohort 1920	0.96379834	0.007228839	0.007070078
Cohort 1925	0.85707828	0.007664768	0.007515334
Cohort 1930	0.68766410	0.008504490	0.008340089
Cohort 1935	0.46431091	0.009646345	0.009444895
Cohort 1940	0.21123472	0.010985080	0.010740653
Cohort 1945	-0.08147459	0.012467199	0.012155399
Cohort 1950	-0.38755739	0.014078607	0.013623592
Cohort 1955	-0.57519153	0.015798917	0.015071518
Cohort 1960	-0.80177901	0.017766331	0.016314446
Cohort 1965	-1.17170766	0.021307055	0.017063635
Cohort 1970	-1.55376457	0.029799279	0.015925719
Cohort 1975	-1.59314497	0.045762410	0.013660624
Cohort 1980	-1.53131132	0.072695827	0.033188400
Cohort 1985	-1.57176249	0.139243354	0.108142241

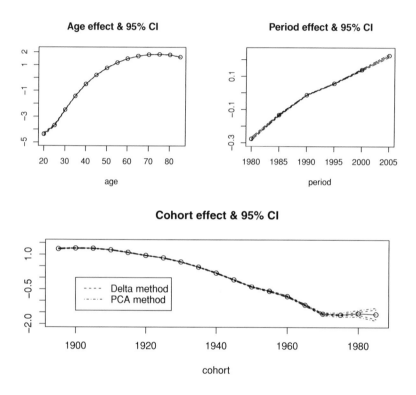

FIGURE 9.2

Plot of the effect estimates and standard errors of the intrinsic estimator of the loglinear model by the maximum likelihood approach with two methods for the standard errors: the PCA and Delta methods.

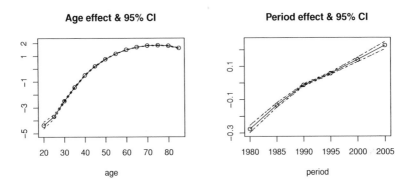

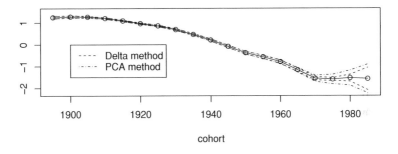

FIGURE 9.3
Plot of the effect estimates and standard errors of the intrinsic estimator of
the loglinear model by the quasi-likelihood approach with two methods for
the standard errors: the PCA and Delta methods.

TABLE 9.2
Comparison of Standard Errors of the Intrinsic Estimator Between the PCA
and Delta Methods with the Loglinear Model Fitted by the Quasi-Likelihood
Approach to the Lung Cancer Mortality Data among US Males

	Estimate	StdErr.PCA	StdErr.Delta
Intercept	-8.11098305	0.021501830	0.021501830
Age 20	-4.35991764	0.133811881	0.133811881
Age 25	-3.68300089	0.082631707	0.082631707
Age 30	-2.48108478	0.049035812	0.049035812
Age 35	-1.41191334	0.034706663	0.034706663
Age 40	-0.49049301	0.027184229	0.027184229
Age 45	0.21951963	0.022206962	0.022206962
Age 50	0.77307172	0.018133549	0.018133549
Age 55	1.18864017	0.014631868	0.014631868
Age 60	1.49133295	0.011940491	0.011940491
Age 65	1.69547097	0.010620734	0.010620734
Age 70	1.81122490	0.011145996	0.011145996
Age 75	1.83502319	0.013306728	0.013306728
Age 80	1.79487989	0.016528171	0.016528171
Age 85	1.61724625	0.020394548	0.020394548
Period 1980	-0.27591158	0.011569866	0.011054458
Period 1985	-0.13278598	0.007403819	0.006586756
Period 1990	-0.01289789	0.003939049	0.002152015
Period 1995	0.05637571	0.004008070	0.002285445
Period 2000	0.14006679	0.007405253	0.006633891
Period 2005	0.22515294	0.011379839	0.010879419

```
             Estimate   StdErr.PCA StdErr.Delta
Cohort 1895  1.25046113 0.031685649 0.026834930
Cohort 1900  1.27564055 0.025252535 0.023279651
Cohort 1905  1.26564590 0.021433545 0.020437332
Cohort 1910  1.20092553 0.018937364 0.018314138
Cohort 1915  1.09093409 0.017591812 0.017122347
Cohort 1920  0.96379834 0.017415660 0.017033175
Cohort 1925  0.85707828 0.018465897 0.018105882
Cohort 1930  0.68766410 0.020488948 0.020092875
Cohort 1935  0.46431091 0.023239897 0.022754564
Cohort 1940  0.21123472 0.026465166 0.025876294
Cohort 1945 -0.08147459 0.030035874 0.029284689
Cohort 1950 -0.38755739 0.033918065 0.032821848
Cohort 1955 -0.57519153 0.038062623 0.036310179
Cohort 1960 -0.80177901 0.042802502 0.039304631
Cohort 1965 -1.17170766 0.051332784 0.041109571
Cohort 1970 -1.55376457 0.071792182 0.038368113
Cohort 1975 -1.59314497 0.110250428 0.032911065
Cohort 1980 -1.53131132 0.175138198 0.079957225
Cohort 1985 -1.57176249 0.335463963 0.260535413
```

9.2   Selection of Side Conditions

As a special case of the analysis of variance (ANOVA) models, the APC models require side conditions for the age, period, and cohort effect estimation. In general, there are two types of side conditions in the ANOVA models. I demonstrate them using the one-way ANOVA model

$$Y_{ij} = \mu + \alpha_i + \varepsilon_{ij} \text{ with } \varepsilon_{ij} \sim iid\, N(0, \sigma^2), \quad i = 1, \dots, a;\, j = 1, \dots, n_i. \quad (9.2)$$

One type of side condition is the centralization in equation (9.3)

$$\sum_{i=1}^{a} \alpha_i = 0, \quad (9.3)$$

and the other type is setting a reference level in equation (9.4)

$$\alpha_l = 0 \quad \text{for some } l \ \ 1 \le l \le a. \quad (9.4)$$

These side conditions have been frequently used in practice ever since the ANOVA models were developed. Often investigators choose side conditions according to their practical needs, such as choosing the white population as the reference level in comparing family income among racial groups. However, theoretical justification is rarely provided. There has been virtually no work done in this area on side condition selection. I point out that selecting side conditions has major impact on estimation of model parameters, in particular, on trend estimation in APC analysis.

I now study the selection of side conditions for the one-way ANOVA, two-way ANOVA, and APC models.

9.2.1 Side Conditions for One-Way ANOVA Models

For convenience, I will use the traditional notations for the sums and means of the quantities.

$$Y_{i.} = \sum_j Y_{ij}, \quad Y_{.j} = \sum_i Y_{ij}, \quad Y_{..} = \sum_i \sum_j Y_{ij},$$

for the sums, and

$$\bar{Y}_{i.} = \frac{1}{n_i} \sum_j Y_{ij}, \quad \bar{Y}_{..} = \frac{1}{n} \sum_i \sum_j Y_{ij},$$

for the means, where $n = \sum_i n_i$.

Consider the estimates of the one-way ANOVA model (9.2) parameters $\mu, \alpha_1, \ldots, \alpha_a$. Let $L = \sum_{i=1}^a c_i \alpha_i$ be a linear combination of the parameters. I examine their expected values and variances of the parameter estimates $\widehat{\alpha}_i$ and the linear combinations $\widehat{L} = \sum_{i=1}^a c_i \widehat{\alpha}_i$ under different side conditions.

1. Setting a reference level $\alpha_l = 0$ for some $1 \le l \le a$

The least-squares method minimizes the residual sum of squares $RSS = \sum_i \sum_j (Y_{ij} - \mu - \alpha_i)^2$ and yields the following parameter estimates $\widehat{\mu} = \bar{Y}_{l.}$, and $\widehat{\alpha}_i = \bar{Y}_{i.} - \bar{Y}_{l.}$ for $i \neq l$. It is straightforward to verify that they are unbiased estimates. $E(\widehat{\mu}) = E(\bar{Y}_{l.}) = \mu$. $E(\widehat{\alpha}_i) = E(\bar{Y}_{i.} - \bar{Y}_{l.}) = \alpha_i$. The linear combination becomes $L = \sum_{i \neq l} c_i \alpha_i$ as $\alpha_l = 0$ and its estimate has the expected value $E(\widehat{L}) = \sum_{i \neq l} c_i E(\widehat{\alpha}_i) = \sum_{i \neq l} c_i \alpha_i = L$. The variances are

$$Var(\widehat{\mu}) = Var(\bar{Y}_{l.}) = \frac{1}{n_l} \sigma^2,$$

$$Var(\widehat{\alpha}_i) = Var(\bar{Y}_{i.} - \bar{Y}_{l.}) = \left(\frac{1}{n_i} + \frac{1}{n_l} \right) \sigma^2 \text{ for } i \neq l.$$

Thus

$$
\begin{aligned}
Var(\widehat{L}) &= Var\left[\sum_{i\neq l} c_i \widehat{\alpha}_i\right] = Var\left[\sum_{i\neq l} c_i(\overline{Y}_{i\cdot} - \overline{Y}_{l\cdot})\right] \\
&= Var\left[\left(\sum_{i\neq l} c_i \overline{Y}_{i\cdot}\right) - \left(\sum_{i\neq l} c_i\right)\overline{Y}_{l\cdot}\right] \\
&= \sum_{i\neq l} c_i^2 \frac{\sigma^2}{n_i} + \left(\sum_{i\neq l} c_i\right)^2 \frac{\sigma^2}{n_l} = \sum_{i=1}^{a} c_i^2 \frac{\sigma^2}{n_i} + \left[\left(\sum_{i\neq l} c_i\right)^2 - c_l^2\right]\frac{\sigma^2}{n_l} \\
&= \sum_{i=1}^{a} c_i^2 \frac{\sigma^2}{n_i} + \left(\sum_{i=1}^{a} c_i\right)\left(\sum_{i=1}^{a} c_i - 2c_l\right)\frac{\sigma^2}{n_l}
\end{aligned}
$$

Hence, for a contrast $L = \sum_i c_i \alpha_i$, since $\sum_i c_i = 0$

$$
Var(\widehat{L}) = \sum_{i=1}^{a} c_i^2 \frac{\sigma^2}{n_i}.
$$

2. A general centralization $\sum_{i=1}^{a} n_i \alpha_i = 0$

The least-squares under a general centralization condition

$$
\sum_{i=1}^{a} n_i \alpha_i = 0, \tag{9.5}
$$

leads to different parameter estimates as follow. $\widehat{\mu} = \overline{Y}_{\cdot\cdot}$ and $\widehat{\alpha}_i = \overline{Y}_{i\cdot} - \overline{Y}_{\cdot\cdot}$. It is straightforward to verify that they are unbiased estimates, i.e. $E(\widehat{\mu}) = \mu$ and $E(\widehat{\alpha}_i) = \alpha_i$. Thus, a linear combination $\widehat{L} = \sum c_i \widehat{\alpha}_i$ has the expectation $E(\widehat{L}) = \sum_i c_i E(\widehat{\alpha}_i) = \sum_i c_i \alpha_i = L$, so $\widehat{L}$ is also unbiased. Now for the variances,

$$
Var(\widehat{\mu}) = Var(\overline{Y}_{\cdot\cdot}) = \frac{\sigma^2}{n},
$$

$$Var(\widehat{\alpha}_l) = Var(\overline{Y}_{l\cdot} - \overline{Y}_{\cdot\cdot}) = Var\left(\frac{1}{n_l}\sum_j Y_{lj} - \frac{1}{n}\sum_i\sum_j Y_{ij}\right)$$

$$= Var\left[\left(\frac{1}{n_l} - \frac{1}{n}\right)\sum_j Y_{lj} - \frac{1}{n}\sum_{i\neq l}\sum_j Y_{ij}\right]$$

$$= \left(\frac{1}{n_l} - \frac{1}{n}\right)^2 n_l\sigma^2 + \frac{1}{n^2}\sum_{i\neq l} n_i\sigma^2$$

$$= \left(\frac{1}{n_l} - \frac{1}{n}\right)\sigma^2 \text{ for } 1 \leq l \leq a.$$

Thus, for each $i = 1, \ldots, a$,

$$Var(\widehat{\alpha}_i) = \left(\frac{1}{n_i} - \frac{1}{n}\right)\sigma^2.$$

A special case of a balanced design, where all groups have an equal number of observations, the general centralization condition (9.5) becomes a regular centralization (9.3) with the number of observations n_0 in each group and the total number of observations $n = an_0$. Then, the variances become

$$Var(\widehat{\mu}) = \frac{\sigma^2}{n}, \quad Var(\widehat{\alpha}_i) = (\frac{1}{n_0} - \frac{1}{n})\sigma^2 = \frac{a-1}{n}\sigma^2 \text{ for all } 1 \leq i \leq a.$$

3. A regular centralization $\sum_{i=1}^a \alpha_i = 0$

The least-squares under the regular centralization condition (9.3) with unequal numbers of observations yields the parameter estimates as follows.

$$\widehat{\mu} = \frac{1}{a}\sum_i \overline{Y}_{i\cdot}$$

and

$$\widehat{\alpha}_i = \left(1 - \frac{1}{a}\right)\overline{Y}_{i\cdot} - \frac{1}{a}\sum_{l\neq i}\overline{Y}_{l\cdot} \text{ for } i = 1, \ldots, a.$$

It is straightforward to verify that they are unbiased, i.e.

$$E(\widehat{\mu}) = \frac{1}{a}\sum_i E(\overline{Y}_{i\cdot}) = \frac{1}{a}\sum_i(\mu + \alpha_i) = \mu + \frac{1}{a}\sum_i \alpha_i = \mu,$$

$$
\begin{aligned}
E(\widehat{\alpha}_i) &= \left(1 - \frac{1}{a}\right) E\overline{Y}_{i\cdot} - \frac{1}{a} \sum_{l \neq i} E\overline{Y}_{l\cdot} \\
&= \left(1 - \frac{1}{a}\right)(\mu + \alpha_i) - \frac{1}{a} \sum_{l \neq i}(\mu + \alpha_l) \\
&= \alpha_i - \frac{1}{a} \sum \alpha_l = \alpha_i.
\end{aligned}
$$

Their variances are

$$
Var(\widehat{\mu}) = Var\left(\frac{1}{a} \sum_i \overline{Y}_{i\cdot}\right) = \frac{1}{a^2} \sum_i \frac{1}{n_i}\sigma^2 = \frac{\sigma^2}{a^2} \sum \frac{1}{n_i},
$$

$$
Var(\widehat{\alpha}_i) = \left(1 - \frac{1}{a}\right)^2 \frac{\sigma^2}{n_i} + \frac{1}{a^2} \sum_{l \neq i} \frac{\sigma^2}{n_l} = \frac{1}{n_i}\sigma^2 + \frac{1}{a^2} \sum_{i=1}^{a} \frac{1}{n_i}\sigma^2 - \frac{2}{an_i}\sigma^2.
$$

Comparison of the variances between different side conditions

Based on the above calculations, the parameter estimates $\widehat{\mu}$ and $\widehat{\alpha}_i$ have different variances under different side conditions. For any given unbalanced data with unequal n_i in each group, the general centralization (9.5) yields the smallest variance for both parameter estimate $\widehat{\mu}$ and $\widehat{\alpha}_i$ with variances

$$
Var(\widehat{\mu}) = \frac{\sigma^2}{n},
$$

$$
Var(\widehat{\alpha}_i) = \left(\frac{1}{n_i} - \frac{1}{n}\right)\sigma^2.
$$

Hence, for unbalanced data, the general centralization condition (9.5) is recommended for efficient estimation of the parameters. For balanced data, the general centralization condition becomes the regular centralization (9.5), and thus the centralization condition (9.3) is recommended for efficient parameter estimation. This applies to any general parameter estimation. However, for the special case of contrasts, the variance remains the same, as shown below.

Under the general centralization side condition (9.5), the estimate of a contrast $\widehat{L} = \sum c_i \widehat{\alpha}_i = \sum c_i (\overline{Y}_{i\cdot} - \overline{Y}_{..}) = \sum c_i \overline{Y}_{i\cdot}$. It has the variance

$$
Var(\widehat{L}) = \sum c_i^2 \frac{\sigma^2}{n_i}.
$$

Under the regular centralization condition (9.3), the estimate of a contrast

$$\widehat{L} = \sum c_i \widehat{\alpha}_i = \sum c_i \left[\left(1 - \frac{1}{a} \right) \overline{Y}_{i\cdot} - \frac{1}{a} \sum_{l \neq i} \overline{Y}_{l\cdot} \right]$$

$$= \sum c_i \overline{Y}_{i\cdot} - \frac{1}{a} \sum c_i \sum_l \overline{Y}_{l\cdot} = \sum c_i \overline{Y}_{i\cdot}.$$

It has the variance

$$Var(\widehat{L}) = \sum c_i^2 \frac{\sigma^2}{n_i} .$$

Hence, a contrast has the same variance under different side conditions, whether through setting a reference level, a general centralization, or a regular centralization. This is a special feature of contrasts, but is not true for a general linear combination. Overall, the general centralization condition is preferred in estimating a linear combination, because its parameter estimates have the smallest variance in comparison with the other side conditions. If the study has balanced data, the regular centralization is preferred.

9.2.2 Side Conditions for Two-Way ANOVA Models

Similar to the side conditions for the above one-way ANOVA models, two kinds of side conditions apply to the two-way ANOVA models with no interaction;

$$Y_{ijk} = \mu + \alpha_i + \beta_j + \varepsilon_{ijk} \text{ with } \varepsilon_{ijk} \sim iid\, N(0, \sigma^2), \tag{9.6}$$

with $i = 1, \ldots, a; \; j = 1, \ldots, p; \; k = 1, \ldots, n_{ij}$ either with centralizations

$$\sum_{i=1}^{a} n_{i\cdot} \alpha_i = 0 \text{ and } \sum_{j=1}^{p} n_{\cdot j} \beta_j = 0 , \tag{9.7}$$

or by setting reference levels, as in equation (9.8)

$$\alpha_{l_1} = 0 \text{ and } \beta_{l_2} = 0 \text{ for some } 1 \leq l_1 \leq a \text{ and } 1 \leq l_2 \leq p. \tag{9.8}$$

To make the discussion simple and easy to understand, I do not consider the combination of centralization on one variable and reference level on the other, but rather take simultaneous centralization or reference levels on both variables. Furthermore, I only consider the balanced design with the same number of observations in each cell (i, j), i.e., n_{ij}. is a constant and does not vary with either i or j.

Using similar techniques and arguments as in the one-way ANOVA models, I summarize the results for the two-way ANOVA models as follows. Interested readers may complete the exercises at the end of this chapter to confirm the results for the two-way ANOVA models.

1. Reference level side conditions $\alpha_{l_1} = 0$ and $\beta_{l_2} = 0$

The following least-squares estimates are unbiased.

$$\widehat{\mu} = \overline{Y}_{l_1 \cdot \cdot} + \overline{Y}_{\cdot l_2 \cdot} - \overline{Y}_{\cdot \cdot \cdot}$$

$$\widehat{\alpha}_i = \overline{Y}_{i \cdot \cdot} - \overline{Y}_{l_1 \cdot \cdot} \text{ for } i \neq l_1$$

$$\widehat{\beta}_j = \overline{Y}_{\cdot j \cdot} - \overline{Y}_{\cdot l_2 \cdot} \text{ for } j \neq l_2$$

The linear combinations $\widehat{L_1} = \sum_i c_i \widehat{\alpha}_i$ and $\widehat{L_2} = \sum_j d_j \widehat{\beta}_j$ are thus unbiased as well. Through some algebra, the variances are

$$Var(\widehat{\mu}) = Var(\overline{Y}_{l_1 \cdot \cdot} + \overline{Y}_{\cdot l_2 \cdot} - \overline{Y}_{\cdot \cdot \cdot}) = \frac{a + p - 1}{n}\sigma^2$$

$$Var(\widehat{\alpha}_i) = \frac{2}{pK}\sigma^2, \text{ for } i \neq l_1$$

$$Var(\widehat{\beta}_j) = \frac{2}{aK}\sigma^2, \text{ for } j \neq l_2$$

$$
\begin{aligned}
Var[\widehat{L_1}] &= Var\left[\sum_i c_i(\overline{Y}_{i \cdot \cdot} - \overline{Y}_{l_1 \cdot \cdot})\right] = Var\left[\sum c_i \overline{Y}_{i \cdot \cdot} - \left(\sum c_i\right)\overline{Y}_{l_1 \cdot \cdot}\right] \\
&= \sum \frac{c_i^2}{n_{i \cdot \cdot}}\sigma^2 + \frac{(\sum c_i)^2}{n_{l_1 \cdot \cdot}}\sigma^2 - \frac{2c_{l_1}(\sum c_i)}{n_{l_1 \cdot \cdot}}\sigma^2 .
\end{aligned}
$$

Similarly,

$$
\begin{aligned}
Var[\widehat{L_2}] &= Var\left[\sum_j d_j(\overline{Y}_{\cdot j\cdot} - \overline{Y}_{\cdot l_2 \cdot})\right] \\
&= \sum \frac{d_j^2}{n_{\cdot j\cdot}}\sigma^2 + \frac{(\sum d_j)^2}{n_{\cdot l_2\cdot}}\sigma^2 - \frac{2d_{l_2}(\sum d_i)}{n_{\cdot l_2\cdot}}\sigma^2.
\end{aligned}
$$

Hence, if L_1 and L_2 are contrasts, their variances are

$$
Var[\widehat{L_1}] = \sum \frac{c_i^2}{n_{i\cdot\cdot}}\sigma^2,
$$

and

$$
Var[\widehat{L_2}] = \sum \frac{d_j^2}{n_{\cdot j\cdot}}\sigma^2.
$$

2. A centralization condition $\sum \alpha_i = 0$ and $\sum \beta_j = 0$

Similarly, it is straightforward to prove that the following least-squares estimates are unbiased.

$$
\widehat{\mu} = \overline{Y}_{\ldots}
$$

$$
\widehat{\alpha}_i = \overline{Y}_{i\cdot\cdot} - \overline{Y}_{\ldots},
$$

$$
\widehat{\beta}_j = \overline{Y}_{\cdot j\cdot} - \overline{Y}_{\ldots}.
$$

Thus, the linear combinations $\widehat{L_1} = \sum_i c_i\widehat{\alpha}_i$ and $\widehat{L_2} = \sum_j d_j\widehat{\beta}_j$ are unbiased as well. Following some algebra, the variances are

$$
Var(\widehat{\mu}) = Var(\overline{Y}_{\ldots}) = \frac{1}{n}\sigma^2
$$

$$
Var(\widehat{\alpha}_i) = \left(\frac{1}{pK} - \frac{1}{n}\right)\sigma^2, \text{ for } i \neq l_1
$$

$$
Var(\widehat{\beta}_j) = \left(\frac{1}{aK} - \frac{1}{n}\right)\sigma^2, \text{ for } j \neq l_2
$$

$$
Var[\widehat{L_1}] = Var\left[\sum_i c_i(\overline{Y}_{i\cdot\cdot} - \overline{Y}_{\ldots})\right] = \sum\left[c_1 - \frac{\sum c_i}{a}\right]^2 \frac{1}{n_{i\cdot\cdot}}\sigma^2
$$

Similarly,

$$Var[\widehat{L_2}] \;=\; Var\left[\sum_j d_j(\overline{Y}_{\cdot j \cdot} - \overline{Y}_{\cdots})\right] = \sum\left[d_1 - \frac{\sum d_i}{a}\right]^2 \frac{1}{n_{\cdot j \cdot}}\sigma^2$$

Hence, if L_1 and L_2 are contrasts, their variances are

$$Var[\widehat{L_1}] = \sum \frac{c_i^2}{n_{i\cdot\cdot}}\sigma^2 \,,$$

and

$$Var[\widehat{L_2}] = \sum \frac{d_j^2}{n_{\cdot j\cdot}}\sigma^2 \,,$$

and are equal to the variances under the side conditions of reference levels.

3. Comparison of the variances between different side conditions

Based on the above calculation, the estimates of parameters μ, α_i and β_j have different variances under the side condition of setting reference levels (9.8) from using centralization (9.7). For any given balanced data with equal numbers of observations in each cell (i, j), the centralization (9.7) yields the smallest variance for all parameter estimates $\widehat{\mu}, \widehat{\alpha}_i$ and $\widehat{\beta}_j$, thus achieving the most efficient estimation. For linear combinations of the parameters, it may be implied that the centralization (9.7) yield the most efficient estimation as well. However, in a special case of contrasts, the variance of the contrasts remain constant as shown in the above, which is a good feature of contrasts, because contrasts are estimable functions and thus remain constant regardless of side conditions.

9.2.3   Side Conditions for Age-Period-Cohort Models

The APC models (4.4) are special cases of the two-way ANOVA models with the fixed cohort effects as special interaction terms. While the age groups have the same number of p observations and the periods have the same number of a observations, the cohort groups have a varying number of observations, from 1 observation on the two extreme cohorts to the largest number of observations $(\min(a,p))$ on the central cohorts. The selection of side conditions on the parameters α_i, β_j and γ_k plays a critical role in parameter estimation. For efficient estimation, it would be straightforward to follow the balanced data design to select the regular centralization conditions (9.7) for the age and period effects α_i and β_j to achieve the most efficient estimation. I then select this centralization on the age and period effects for all APC models without further specification.

Now consider the side conditions for the cohort effects. Since the number of observations varies with the cohort, the imbalance of the cohort data needs further studying on the side condition selection for the cohort effects. Similar to the above study for one-way ANOVA models, three side conditions can be considered in general: the reference levels

$$\gamma_l = 0 \text{ for some } 1 \le l \le a + p - 1 , \tag{9.9}$$

regular centralization

$$\sum_{k=1}^{a+p-1} \gamma_k = 0 , \tag{9.10}$$

and general centralization

$$\sum_{k=1}^{a+p-1} n_k \gamma_k = 0 , \quad k = a - i + j , \tag{9.11}$$

where n_k is the number of observations on the kth diagonal. Since general centralization is not as popular as regular centralization in APC analysis, I will focus on regular centralization, and compare it with reference levels.

The APC models are complex and it is difficult to derive a closed form for the parameter estimates; see the complex estimates in the reduced age-cohort model in question 7 of Chapter 3 exercises. For this reason, I will present some simulation results on the variance of the cohort effect estimates in comparing between the reference levels and the regular centralization (9.10).

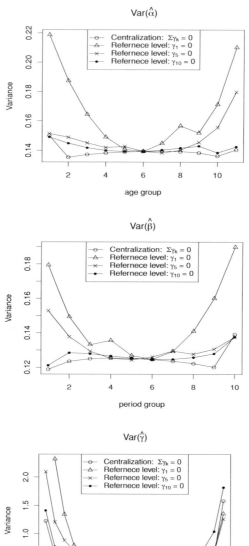

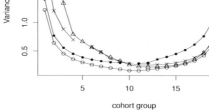

FIGURE 9.4
Comparison of variance of age, period, and cohort effect estimates by 1000
simulation runs with varying side conditions on cohort effects. Regular centralization conditions are specified on age and period effects.

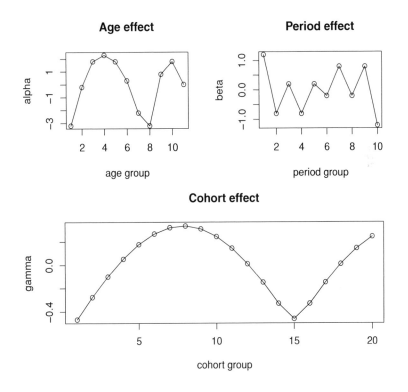

FIGURE 9.5

Plot of age, period, and cohort effects specified for the simulation study to examine the effect of varying side conditions on the variance of age, period and cohort effect estimates.

Figure 9.4 presents a comparison of the variance of age, period, and cohort effect estimates by a simulation study of 11 age effects, 10 period effects, and 20 cohort effects specified as shown in Figure 9.5. In the estimation of age, period, and cohort trends, the intrinsic estimator defined by the unique estimator in the non-null parameter space and perpendicular to the null space is computed for each APC data set with specified side conditions on the cohort effects, including the regular centralization (9.10), and three reference levels: an extreme cohort $\gamma_1 = 0$ with 1 observation on the cohort, an intermediate cohort $\gamma_5 = 0$ with 5 observations on the cohort, or a central cohort $\gamma_{10} = 0$ with the maximum of 10 observations on the cohort. It is shown that the reference levels present larger variances than the centralization condition. Furthermore, the reference level on the extreme cohort effect presents the largest variance while the reference level on the central cohort effect presents smaller variances than the reference level on the extreme and the intermediate cohort effects. For the presentation of the variance, the 0 variance of the selected reference level (γ_1, γ_5, or γ_{10}) is dropped from the plot of the cohort effect variance in Figure 9.4. Interested readers may find more simulation results and applications to real APC data sets for comparison of different side conditions in Fu et al (2017).

9.2.4   Conclusion on Side Condition Selection

Based on the above theoretical derivation and simulation results, I conclude that the regular centralization condition is recommended for one-way ANOVA, two-way ANOVA, and APC models as an efficient side condition for parameter estimation. In particular, the centralization condition in APC models not only makes parameter estimation efficient, thus achieving high accuracy, but also helps to make trend estimation sensible. Notice that confidence intervals of the estimated trends in the age, period, and cohort help to present the variability due to sampling uncertainty in the data. However, such uncertainty would be incorrectly presented by a trend going through a specified reference level 0 with no variation, which is misleading and causes misunderstanding and misinterpretation, as discussed by Kupper et al (1985).

In summary, the intrinsic estimator of APC models yields the most efficient estimation and presents the best analysis results when the regular centralization conditions are specified for age, period, and cohort effects.

9.3 Suggested Readings

The Delta method is a popular tool for calculating the variance of nonlinear functions of a random variable, and often yields efficient estimation of variance. It appears in many textbooks, for example, Sen and Srivastava (1990).

9.4 Exercises

1.* Based on the Delta method for the variance estimation of the period and cohort effects, discuss why in the loglinear models, the two factor age-cohort model is preferred to the single factor cohort model.

---

* Difficult exercises with an asterisk are meant for graduate students in biostatistics.

10

Unequal Spans in Age and Period Groups with Applications to Survey Data

So far, I have studied age-period-cohort data with identical spans in age and period groups, such as 5 years in both. Such identical spans make the data fit into the Lexis diagram and the corresponding APC models have well-defined birth cohort effects. However, in many studies, especially in social studies with survey data, it is often not practical to expect identical spans in age and period groups, because the survey needs to be conducted frequently to promptly address the needs. Sometimes, the total period of time across several surveys in the study is relatively short, not long enough to meet the requirement of the age-period-cohort models. For example, the survey data on the US family retirement accounts in Table 1.4 has age-group spans of 10 years, but the periods in which the eight surveys were conducted are 3 years apart from 1989 to 2010. The regular age-period-cohort models for identical spans would require regrouping the periods into 10 year spans, and only two periods of 10 full years are covered in the study, not enough to fit a regular APC model. Furthermore, because of the relatively short period of time of the study, the period trend would not provide enough useful information if multiple periods are collapsed into 10 year spans. Hence, novel APC models with unequal spans need to be developed so that age, period, and cohort trends can be accurately estimated. This chapter studies APC models for unequal span data.

10.1   APC Data with Unequal Spans

Most APC studies in the literature have identical time spans in age and period groups, such as 5 years in most public health studies, or 10 years in demography or economic studies. These identical time spans make it easy to identify birth cohorts by the diagonals of the table since the individuals on the same diagonal were born in about the same years and thus belong to the same birth cohort, as discussed in Section 4.1. When the age and period groups have unequal time spans, however, it becomes difficult to identify birth cohorts because the diagonals of the table do not represent complete birth cohorts, as shown in the retirement accounts data in Table 1.4 with 10 year age-group span and 3 year period span. Identifying complete cohorts is important since it is the effects of birth cohorts that is of interest in the APC studies, but not the effects of the diagonals of the table, which present spurious effects in unequal span data.

The following collapsing strategy helps to correctly identify birth cohorts in the retirement accounts data. Notice that three periods together span about the same number of years as one age group; a table may be built to have the same age groups but different periods by collapsing three consecutive periods into one. Then the newly formed table would have birth cohorts on the diagonals, but have fewer periods. While such collapsing strategy correctly identifies birth cohorts, the resulting fewer number of periods will incur loss of period information. To compensate the loss, an intend-to-collapse (ITC) method has been recently developed (Fu 2018), which allows correct identification of birth cohorts while retaining the original age groups and periods so that no information in age and period will be lost.

TABLE 10.1
Birth Cohort Structure of the Retirement Accounts Data

	Year							
Age	1989	1992	1995	1998	2001	2004	2007	2010
25–34	coh 6	coh 7	coh 7	coh 7	coh 8	coh 8	coh 8	coh 9
35–44	coh 5	coh 6	coh 6	coh 6	coh 7	coh 7	coh 7	coh 8
45–54	coh 4	coh 5	coh 5	coh 5	coh 6	coh 6	coh 6	coh 7
55–64	coh 3	coh 4	coh 4	coh 4	coh 5	coh 5	coh 5	coh 6
65–74	coh 2	coh 3	coh 3	coh 3	coh 4	coh 4	coh 4	coh 5
75+	coh 1	coh 2	coh 2	coh 2	coh 3	coh 3	coh 3	coh 4

10.2   The Intend-to-Collapse (ITC) Method

The ITC method takes the APC unequal span data, and rearranges it by tentatively collapsing multiple consecutive periods into one if multiple periods span the same duration as one age span, or collapsing multiple consecutive age groups into one if multiple age groups span the same duration as one period span, so that the age groups and periods would have identical spans, and consequently, the diagonals of the table would identify complete birth cohorts. Notice that the above collapsing is only tentative to allow correct identification of birth cohorts, but the age groups and periods will not be collapsed. After correctly coding the cohort effects, the age and period effects will be coded according to the original table. In such a way, the age groups and periods will remain unchanged to retain the same information while the cohort effects will be modeled correctly following true birth cohorts rather than the nominal diagonals of the table.

In the following, I illustrate the ITC method with the retirement accounts data in Table 1.4. The data has 6 age groups from < 35, 35–44, to 65–74 and 75+, and 8 periods from 1989, 1992 to 2007 and 2010. For technical simplicity, the first age group will be regarded as 25–34 and the last one as 75–84 so that the related cohorts will be easy to define. Since the periods are 3 years apart, three consecutive ones span about 10 years, such as from 1990 to 1999, and from 2000 to 2009. In tentatively collapsing them, the 3 consecutive periods in 1992, 1995 and 1998 would be collapsed into one, and so would the periods 2001, 2004 and 2007. Hence, eight original periods would be collapsed into four 10-year periods with the first (1989) and the last (2010) as single year representatives of their corresponding periods. This will allow correct coding of the cohort effects according to birth cohorts rather than following the diagonals of the table. Table 10.1 shows the cohort structure and the coding by the ITC method. Notice that the ITC method does not alter the coding of the age and period effects.

10.3   APC Models for Unequal Spans

The above coding by the ITC method makes it possible to fit the age-period-cohort linear model

$$Y_{ij} = \mu + \alpha_i + \beta_j + \gamma_k + \varepsilon_{ij}, \ i = 1, \ldots, a; \ j = 1, \ldots, p, \qquad (10.1)$$

where the model parameters μ, α_i, β_j and the random error term ε_{ij} remain the same as in the linear APC models (4.4). γ_k is the cohort effect of the kth birth cohort coded by the ITC method, starting from γ_1 for the oldest cohort in the study to the youngest cohort. Notice that the previous index relationship $k = a - i + j$ in the equal span model (4.4) is not true anymore, and the number of cohorts may vary, depending on how the ITC approach "collapses" the periods. Often, it is preferred to have the most possible cohorts in the model by making the first and last periods incomplete so that more cohort information can be obtained in the analysis, as shown in the example of the retirement accounts data in Table 10.1.

Similarly, the generalized linear models may also be fitted to unequal span data.

$$g(EY_{ij}) = \mu + \alpha_i + \beta_j + \gamma_k \,, \ i = 1, \ldots, a; \ j = 1, \ldots, p \,, \qquad (10.2)$$

where the response variable Y_{ij} follows an exponential family distribution, and the model parameters μ, α_i, β_j and γ_k remain the same as in the linear model (10.1). In particular, the loglinear model with response variable following the Poisson distribution is of interest in APC studies.

In the next section, I will study the identification problem and the model fitting method of the APC models (10.1) and (10.2).

10.4   Identification Problem and Intrinsic Estimator for Unequal Span Data

10.4.1   Multiple Estimators and Identification Problem

The above APC models encounter the same identification problem, induced by the linear relation among the age, period, and cohort as discussed in previous chapters. Notice that the relationship among age, period, and cohort indices, i, j and k, respectively, is not the key to the identification problem, but rather the linear relationship among the age, period, and cohort, which exists in the unequal span data as well. A close examination of the design matrix of the unequal span models (10.1) and (10.2) reveals that there exists one and only one eigen value 0, thus the model parameter space has a one-dimensional null space, and the model multiple estimators have exactly the same structural

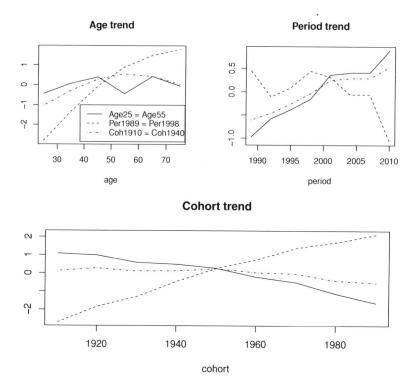

FIGURE 10.1
Plot of three sets of age, period, and cohort trends by linear model (10.1) fitted
to the log-transformed mean value of the retirement accounts with equality
constraints on age $\alpha_{25} = \alpha_{55}$, period $\beta_{1989} = \beta_{1998}$, and cohort $\gamma_{1910} = \gamma_{1940}$.

decomposition as in equation (6.1) for the equal span models (4.4). It thus
induces the same parameter identification problem.

I illustrate the multiple estimators with the retirement accounts data. Figure 10.1 presents three curves of age, period, and cohort trends by linear model
(10.1) fitted to the log-transformed mean value of retirement accounts. They
are uniquely estimated by an equality constraint on the age, period, or cohort
effect, respectively.

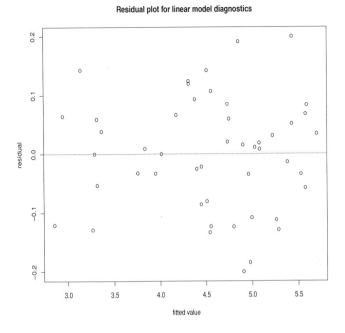

FIGURE 10.2
Residual plot against fitted value for linear model diagnostics on the log-transformed mean value of retirement accounts.

10.4.2 The Intrinsic Estimator for Unequal Span Data

Since the design matrix of the linear models (10.1) has a unique eigen value 0 and the model parameter estimators follow the same decomposition (6.1), the unique projection of the multiple estimators to the non-null space defines the intrinsic estimator of model (10.1) in the same way as for the equal span models (4.4). The intrinsic estimator defined by the projection enjoys the same asymptotic properties as the intrinsic estimator for equal spans. I summarize the results in the following theorem, and leave the proof for the readers as an interesting exercise.

TABLE 10.2

Output of Linear APC Model by the Intrinsic Estimator with **R** Function
apclinkfit on the US Mean Value of Retirement Accounts

```
> apclinkfit(r=retire, header=T,lam=0, p0=1,k=3, transform="log",
    Plot=T, CIplot=T, panelplot=T, agapyr=10, cgapyr=10, pgapyr=3,
    ModelDiag=T, pplim=c(-1.5,0.7))
$model
lm(formula = as.vector(t(r)) ~ xx - 1)
$scale
[1] "log"
$Rsquare
          Rsquare Adjusted.Rsquare        Model df
        0.9995792        0.9992787     20.0000000
$varcomp
Variance Component        Residual df
        0.1237628         28.0000000
$parameters
                Parameter Standard.Error          t        P-value
intercept      4.45583636     0.02708516 164.5120912 0.000000e+00
Age   25      -1.07271493     0.04700508 -22.8212561 0.000000e+00
Age   35      -0.30603039     0.04307488  -7.1046137 6.932138e-07
Age   45       0.27829415     0.04525652   6.1492612 5.227311e-06
Age   55       0.56205323     0.04544427  12.3679659 7.280843e-13
Age   65       0.45630945     0.04371547  10.4381698 1.530880e-09
Age   75       0.08208850     0.04507404   1.8211926 8.357198e-02
Period 1989  -0.58680560     0.04642221 -12.6406229 4.325429e-13
Period 1992  -0.45035888     0.04805443  -9.3718496 9.310918e-09
Period 1995  -0.25526787     0.04805443  -5.3120572 3.368691e-05
Period 1998  -0.03067956     0.04805443  -0.6384336 5.283793e-01
Period 2001   0.23720379     0.04741245   5.0029850 6.826036e-05
Period 2004   0.28388713     0.04741245   5.9876070 7.443502e-06
Period 2007   0.28760538     0.04741245   6.0660305 1.532480e-06
Period 2010   0.51441561     0.05202508   9.8878392 3.823255e-09
Cohort 1910   0.06706393     0.09355512   0.7168387 4.817619e-01
Cohort 1920   0.22757485     0.06522935   3.4888411 1.622869e-03
Cohort 1930   0.08070417     0.05613659   1.4376394 1.659994e-01
Cohort 1940   0.23462536     0.05494706   4.2700254 3.741743e-04
Cohort 1950   0.27396690     0.05574144   4.9149595 3.499384e-05
Cohort 1960   0.05896157     0.05050704   1.1673931 2.567816e-01
Cohort 1970   0.01395645     0.04644210   0.3005128 7.668876e-01
Cohort 1980  -0.35144247     0.05362349  -6.5538898 4.178798e-07
Cohort 1990  -0.60541075     0.12415918  -4.8760854 9.143285e-05
```

Theorem 10.1 Under regularity conditions similar to the ones specified in Chapter 8,

1) the estimator $\widehat{\boldsymbol{\theta}}_p$ of the intercept and age effects $\boldsymbol{\theta}_p = (\mu, \alpha_1, \ldots, \alpha_{a-1})^T$ by the intrinsic estimator of the APC model (10.2) converges in probability to the true parameters as the number of periods p diverges to infinity, i.e., there exists a finite a-dimensional vector $\boldsymbol{\theta}^{\infty}$

$$\widehat{\boldsymbol{\theta}}_p \longrightarrow_p \boldsymbol{\theta}^{\infty} \quad \text{as } p \to \infty$$

in probability, and

2) the estimator $\widehat{\boldsymbol{\theta}}_p$ follows an approximately normal distribution,

$$\sqrt{p}\left(\widehat{\boldsymbol{\theta}}_p - \boldsymbol{\theta}^{\infty}\right) \longrightarrow_d N\left(\mathbf{0}, C_2^{-1}\right) \quad \text{as } p \to \infty,$$

where C_2 is an $a \times a$-dimensional positive-definite matrix.

Interested readers may see the exercises at the end of the chapter for more interesting results of the intrinsic estimator.

10.4.3   Analysis of Unequal Span Data by Intrinsic Estimator

In the following, I demonstrate the intrinsic estimator method and examine its performance in analyzing APC data with unequal spans using the mean value of retirement accounts data with the linear model (10.1). I have illustrated the identification problem with this data set. I fit the model with the intrinsic estimator method to the log-transformed response of the mean value of retirement accounts.

Table 10.2 displays the output of the linear model (10.1) fitted to the mean value of retirement accounts data, including the R^2, the adjusted R_a^2, the variance component, and degrees of freedom, as well as some technical terms special in the unequal spans data, the number of periods k "collapsed" for birth cohorts, the number of periods p_0 for the first incomplete period,

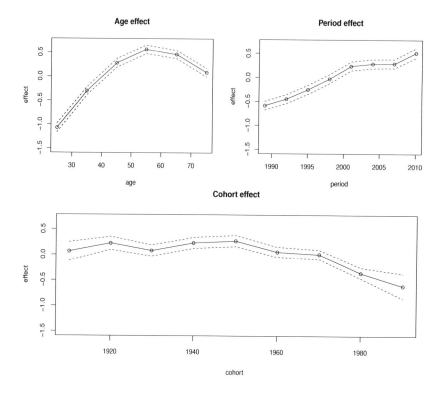

FIGURE 10.3
Plot of the effect estimates and standard errors of the intrinsic estimator of
the linear model for the log-transformed mean value of retirement accounts
with unequal spans.

and the number of birth cohorts K. It also provides parameter estimates of
age, period, and cohort effects, standard errors, the t statistic and the p-value
of each estimate. The model achieves large values of $R^2 = 0.99958$ and the
adjusted $R_a^2 = 0.99928$, indicating a good fit to the data. Figure 10.2 displays
the residual plot against the fitted value for the model diagnostics. It shows
that the residuals are centered around 0 with no deviation, and constant
variation with the fitted values, indicating no violation of the Gauss-Markov
conditions for linear models.

Figure 10.3 shows age, period, and cohort trends and 95% confidence intervals by the parameter estimates. It is shown that the mean value of retirement
accounts increases sharply with the age of the head of household from below
35 to a peak around age 55–64 and then declines rapidly to age 75+. It also
presents a rapidly increasing period trend from 1989 to 2001, followed by a flat
trend from 2001 to 2007 before a sharp increase till 2010. The cohort trend
fluctuates from the oldest cohort born around 1905 to the one around 1950
then decreased from 1950 to the youngest cohort born around 1990.

In summary, the above age, period, and cohort effect estimates lead to
sensible trend estimation and interpretation of the mean value of retirement
accounts of the US families. Following the consistency of the intrinsic estimator, the above age, period, and cohort trend presents accurate estimation of
age, period, and cohort effects.

10.5   Fitting Unequal Span Data with R Function apclinkfit

The APC linear model (10.1) for unequal spans can be fitted using the **R**
function `apclinkfit`, as shown in Table 10.2. The option k = 3 specifies that
3 periods are used in one span in the ITC method. The option p0 = 1 specifies
that the first period in each age group is modeled as an incomplete period sep-
arated from the next k periods for the second period. This option is convenient
to allow the users to choose how many periods are grouped together to form
the first incomplete period so that the most possible birth cohorts may be modeled. The option `transform = "log"` specifies that the log-transformation
is applied to the response. While the log-transformation is the default op-

tion, other option `transform = "id"` can also be specified for identity or no transformation. The option `ModelDiag = T` produces the residual plot for the model diagnostics. Another useful option is the specification `cmin = 1910`, which specifies the mid-birth year of the oldest cohort to be 1910. This is critical since the unequal spans and the incomplete first period option by the command `p0 = 1` allows the specification of the number of periods in the first incomplete period, which opens up the uncertainty of the mid-birth year of the oldest cohort. Hence the option allows the user to calculate the first cohort year using the mid-year of the second cohort and the span to specify an appropriate mid-year of the oldest cohort. The option `pplim = c(-1.5, 0.7)` specifies that the plots of age, period, and cohort trends have a common y-axis range of $(-1.5, 0.7)$. The default is `pplim=c(0,0)`, which allows the plots of age, period, and cohort trends to have their individual ranges.

10.6   Exercises

1. Demonstrate numerically that the design matrix of the APC models (10.1) has one-dimensional null space.

2. * The multiple curves by the equality constraints in Figure 10.1 illustrate the identification problem in the linear model (10.1). As discussed in previous chapters and in the literature, the nonlinear characteristics of the curves are estimable. However, the curves in Figure 10.1 do not seem to possess the same nonlinear characteristics; see the large difference between the age trends, and between the period trends. Can you explain why, and identify what causes

the difference in the nonlinear characteristics? Do you expect this to happen in modeling APC data with equal spans?

3.* Prove that the age effect estimates of the intrinsic estimator of the APC model (10.1) for unequal span data converges in probability to the true parameter values as the number of periods p diverges to infinity. Hence, the intrinsic estimator yields consistent estimation.

4.* Prove that the age effect estimates of the intrinsic estimator of the APC model (10.1) also possess the $\sqrt{p}$-consistency and asymptotic normality.

$$\sqrt{p}\left(\widehat{\boldsymbol{\theta}}_p - \boldsymbol{\theta}^\infty\right) \longrightarrow_d N\left(\mathbf{0}, C_2^{-1}\right) \qquad \text{as } p \to \infty,$$

where C_2 is an $a \times a$-dimensional positive-definite matrix. Assume multiple periods form one span of the same duration as the age span, show that the inverse matrix C_2^{-1} is not the asymptotic variance-covariance matrix of the estimator $\widehat{\boldsymbol{\theta}}_p$, but rather a multiple of the variance-covariance matrix by a constant. What is the constant?

Bibliography

[1] C. Berzuini and D. Clayton. Bayesian-analysis of survival on multiple time scales. *Statistics in Medicine*, 13(8):823–838, Apr 30 1994.

[2] O. Bretscher. *Linear Algebra with Applications*. Prentice Hall, Upper Saddle River, 2004.

[3] D. Clayton and E. Schifflers. Models for temporal variation in cancer rates. I: Age-period and age-cohort models. *Statistics in Medicine*, 6(4):449–467, Jun 1987.

[4] D. Clayton and E. Schifflers. Models for temporal variation in cancer rates. II: Age-period-cohort models. *Statistics in Medicine*, 6(4):469–481, Jun 1987.

[5] S.E. Fienberg. Cohort Analysis' Unholy Quest: A Discussion. *Demography*, 50(6):1981–1984, Dec 2013.

[6] S.E. Fienberg and W.M. Mason. Specification and implementation of age, period, and cohort models. In W.M. Mason and S.E. Fienberg editor, *Cohort Analysis in Social Research — Beyond the Identification Problem*, pages 44–48., 1985.

[7] W.H. Frost. The age selection of mortality from tuberculosis in successive decades. *American Journal of Hygiene*, 30(3):91–96, 1939.

[8] W.J. Fu. Ridge estimator in singular design with application to age-period-cohort analysis of disease rates. *Communications in Statistics – Theory and Methods*, 29(2):263–278, 2000.

[9] W.J. Fu. A smoothing cohort model in age-period-cohort analysis with applications to homicide arrest rates and lung cancer mortality rates. *Sociological Methods & Research*, 36(3):327–361, Feb 2008.

[10] W.J. Fu. Constrained Estimators and Consistency of a Regression Model on a Lexis Diagram. *Journal of American Statistical Association*, 111(513):180–199, Mar 2016.

[11] W.J. Fu. Modeling age-period-cohort data with unequal spans in age and period with applications to survey data, *unpublished manuscript*, 2018.

[12] W.J. Fu, K.K. Gao, and S.J. Huang. Recent advances in age-period-cohort analysis, *unpublished manuscript*, 2017.

[13] W.J. Fu and P. Hall. Asymptotic properties of estimators in age-period-cohort analysis. *Statistics and Probability Letters*, 76(17):1925–1929, Nov 1 2006.

[14] K. Fukuda. Age-period-cohort decomposition of aggregate data: An application to US and Japanese household saving rates. *Journal of Applied Econometrics*, 21(7):981–998, Nov 2006.

[15] B.G. Greenberg, J.J. Wright, and C.G. Sheps. A technique for analyzing some factors affecting the incidence of syphilis. *Journal of American Statistical Association*, 45(251):373–399, 1950.

[16] C. Heuer. Modeling of time trends and interactions in vital rates using restricted regression splines. *Biometrics*, 53(1):161–177, 1997.

[17] T.R. Holford. The estimation of age, period and cohort effects for vital-rates. *Biometrics*, 39(2):311–324, 1983.

[18] T.R. Holford. An alternative approach to statistical age-period-cohort analysis. *Journal of Chronic Diseases*, 38(10):831–836, 1985.

[19] T.R. Holford. Understanding the effects of age, period, and cohort on incidence and mortality-rates. *Annal Review of Public Health*, 12:425–457, 1991.

[20] J.M. Janis, L.L. Kupper, and B.G. Greenberg. Age-period-cohort analysis of lung-cancer mortality data. *Statistics in Medicine*, 114(3):440, 1981.

[21] J.M. Janis, L.L. Kupper, A. Karmous, and B.G. Greenberg. Problems with 2-factor and 3-factor models age-period-cohort analysis. *Biometrics*, 41(1):320, 1985.

[22] I.T. Jolliffe. *Principal Component Analysis*. Springer, New York, 2002.

[23] K. Knight and W.J. Fu. Asymptotics for Lasso-type estimators. *Annals of Statistics*, 28(5):1356–1378, Oct 2000.

[24] R.G. Kuhlen. Social change: A neglected factor in psychological studies of the life span. *School and Society*, 52:14–16, 1940.

[25] L.L. Kupper, J.M. Janis, A. Karmous, and B.G. Greeberg. An alternative approach to statistical age-period-cohort analysis-reply. *Journal of Chronic Diseases*, 38(10):837–840, 1985.

[26] L.L. Kupper, J.M. Janis, A. Karmous, and B.G. Greenberg. Statistical age-period-cohort analysis - a review and critique. *Journal of Chronic Diseases*, 38(10):811–830, 1985.

[27] L.L. Kupper, J.M. Janis, I.A. Salama, and C.N. Yoshizawa. Age-period-cohort analysis — theoretical and philosophical considerations. *Biometrics*, 39(3):806, 1983.

[28] L.L. Kupper, J.M. Janis, I.A. Salama, C.N. Yoshizawa, and B.G. Greenberg. Age-period-cohort analysis—an illustration of the problems in assessing interaction in one observation per cell data. *Communications in Statistics – Theory and Methods*, 12(23):2779–2807, 1983.

[29] W.C. Lee and R.S. Lin. Autoregressive age-period-cohort models. *Statistics in Medicine*, 15(3):273–281, Feb 15 1996.

[30] W.C. Lee and R.S. Lin. Modeling exposure intensity and susceptibility-latency effect from retrospective cohort data - Application to lung cancer mortality in Blackfoot disease endemic area in Taiwan. *Biometrical Journal*, 38(7):843–855, 1996.

[31] W.C. Lee and R.S. Lin. Modelling the age-period-cohort trend surface. *Biometrical Journal*, 38(1):97–106, 1996.

[32] L. Luo, J. Hodges, C. Winship, and D. Powers. The sensitivity of the intrinsic estimator to coding schemes: comment on Yang, Schulhofer-Wohl, Fu, and Land. *American Journal of Sociology*, 122(3):930–961, Nov 2016.

[33] W.M. Mason and B. Entwisle. The age-period-cohort accounting framework and multilevel comparative-analysis. *Biometrics*, 41(1):320–321, 1985.

[34] W.M. Mason, K.O. Mason, and H.H. Winsborough. Cohort analysis futile quest-statistical attempts to separate age, period and cohort effect-reply. *American Sociological Review*, 41(5):904–905, 1976.

[35] W.M. Mason and H.L. Smith. Age-period-cohort analysis and the study of deaths from pulmonary tuberculosis. In W.M. Mason and S.E. Fienberg editor, *Cohort Analysis in Social Research — Beyond the Identification Problem*, pages 151–228., 1985.

[36] P. McCullagh and J.A. Nelder. *Generalized Linear Models*. Chapman & Hall, London, second edition, 1989.

[37] S.A. Murphy and A.W. van der Vaart. On profile likelihood. *Journal of American Statistical Association*, 95(450):449–465, Jun 2000.

[38] J. Neyman and E. Scott. Consistent estimates based on partially consistent observations. *Econometrika*, 16(1):1–32, 1948.

[39] R.M. OBrien. Age period cohort characteristic models. *Social Science Research*, 29(1):123–139, 2000.

[40] C. Osmond. Age, period and cohort models applied to cancer mortality rates. *Statistics in Medicine*, 1(3):245–259, 1982.

[41] C. Robertson and P. Boyle. Age, period and cohort models - the use of individual records. *Statistics in Medicine*, 5(5):527–538, Sep-Oct 1986.

[42] C. Robertson and P. Boyle. Age-period-cohort analysis of chronic disease rates. I: Modelling approach. *Statistics in Medicine*, 17(12):1305–1323, Jun 30 1998.

[43] C. Robertson and P. Boyle. Age-period-cohort models of chronic disease rates. II: Graphical approaches. *Statistics in Medicine*, 17(12):1325–1339, Jun 30 1998.

[44] W.L. Rodgers. Estimable functions of age, period, and cohort effects. *American Sociological Review*, 47(6):774–787, 1982.

[45] W.L. Rodgers. More chimeras of the age-period-cohrt accounting framework – Reply. *American Sociological Review*, 47(6):793–796, 1982.

[46] P.S. Rosenberg and W.F. Anderson. Age-period-cohort models in cancer surveillance research: ready for prime time? *Cancer Epdemiology and Biomarkers & Prevention*, 20(7):1263–1268, Jul 2011.

[47] H. Scheffé. *The Analysis of Variance.* Wiley, New York, 1959.

[48] V. Schmid and L. Held. Bayesian extrapolation of space-time trends in cancer registry data. *Biometrics*, 60(4):1034–1042, Dec 2004.

[49] V.J. Schmid and L. Held. Bayesian age-period-cohort modeling and prediction - BAMP. *Journal of Statistical Software*, 21(8):1–15, Oct 2007.

[50] S. R. Searle. *Linear Models.* Wiley, New York, 1971.

[51] A. Sen and M. Srivastava. *Regression Analysis: Theory, Methods, and Applications.* Springer, New York, 1990.

[52] D.R. Shopland. Tobacco use and its contribution to early cancer mortality with a special emphasis on cigarette smoking. *Environmental Health Perspectives*, 103(8):131–142, Nov 1995.

[53] H.L. Smith. Advances in age-period-cohort analysis. *Sociological Methods & Research*, 36(3):287–296, Feb 2008.

[54] H.L. Smith, W.M. Mason, and S.E. Fienberg. More chimeras of the age-period-cohort accounting framework-comment. *American Sociological Review*, 47(6):787–793, 1982.

[55] R.E. Tarone and K.C. Chu. Implications of birth cohort patterns in interpreting trends in breast-cancer rates. *Journal of the National Cancer Institute*, 84(18):1402–1410, Sep 16 1992.

[56] Y.K. Tu, K.L. Chien, and M. Gilthorpe. Using partial least squares regression for the age-period-cohort analysis. *Journal of Epidemiology and Community Health*, 65(1):A82, Aug 2011.

[57] Y.K. Tu, N. Kraemer, and W-C. Lee. Addressing the identification problem in age-period-cohort analysis: a tutorial on the use of partial least squares and principal components analysis. *Epidemiology*, 23(4):583–593, Jul 2012.

[58] D. Williams. *Probability with Martingales*. Cambridge University Press, New York, 1991.

[59] Y. Yang, W.J. Fu, and K.C. Land. A methodological comparison of age-period-cohort models: The intrinsic estimator and conventional generalized linear models. *Sociological Methodology*, 34:75–110, 2004.

[60] Y. Yang and K.C. Land. A mixed models approach to the age-period-cohort analysis of repeated cross-section surveys, with an application to data on trends in verbal test scores. *Sociological Methodology*, 36:75–97, 2006.

[61] Y. Yang, S. Schulhofer-Wohl, W.J. Fu, and K.C. Land. The intrinsic estimator for age-period-cohort analysis: What it is and how to use it. *American Journal of Sociology*, 113(6):1697–1736, May 2008.

Index